COASTERS 101

An Engineer's Guide to Roller Coaster Design

Nick Weisenberger

Fifth Edition

Coasters 101: An Engineer's Guide to Roller Coaster Design

Contents

Chapter 1: Coasters 101 6
   Brief History of the Roller Coaster 7
   Coaster Classification 9
   Wood Versus Steel 13
   Design Software and Computer Technology 18
Chapter 2: Planning 24
   Design Constraints and Considerations 26
   Selecting a Manufacturer 29
   Layout and Profile 32
   Ride Analysis 33
Chapter 3: Design Principles 35
   G Forces 35
   Kinetic and Potential Energy 40
   Friction and Drag 42
   Aerodynamic Drag 43
   Friction 47
   Wheel Bearing Lubrication 47
   How does temperature affect roller coasters? 48
   How does mass affect roller coasters? 49
   Free Fall Example 51
Chapter 4: Coaster Elements 54
   Lifting Mechanisms 54
   A People-Powered Coaster 64
   Lift Energy Example 65
   Launch Systems 65
   Vertical Loops 73
   Banking the Track Through a Curve 78
      Curve Design Example 81
Chapter 5: Vehicle Design 83
   Seating Configuration 84
   Wheel Design and Material Selection 86
   Restraint Design 93
      Vehicle Design Example 96

  How a 4th Dimension Coaster Works 98
  A Roller Coaster on a Cruise Ship.. 102
  How a Roller Coaster Jumps the Tracks....................................... 108
Chapter 6: Infrastructure 110
  Supports and Foundations... 110
  Brakes, Blocks, Sensors, and Switches 114
  Racing/Dueling Control Systems.. 120
  Safety Systems on the World's Scariest Roller Coaster 122
Chapter 7: From Paper to the Park... 124
  Manufacturing.. 124
  Construction... 127
  Test and Adjust 131
  Operation 134
Chapter 8: Safety 136
  Amusement Industry Safety Standards.. 137
  Lock Out Tag Out Procedures 140
  Inspections and Maintenance... 141
  Why are roller coasters removed? 145
  Four Times a Roller Coaster Saved a Life.................................. 149
Chapter 9: Design Example 151
Chapter 10: Career Advice 155
  Career Advice Summary... 163
Chapter 11: Enjoy the Ride... 164
About the Author 168
The 50 Most *Terrifying* Roller Coasters Ever Built 170
Works by Nick Weisenberger... 171
Appendix I: Acronyms... 172
Appendix II: Glossary.. 174
Appendix III: Notable Milestones in Roller Coaster History 182
Appendix IV: Resources 185

Coasters 101: An Engineer's Guide to Roller Coaster Design

Copyright Information

Fifth Edition – Paperback Version
Copyright ©2022 by Nick Weisenberger
First edition first published in 2012
ISBN-10: 1468013556
ISBN-13: 978-1468013559

All rights reserved. No part of this book may be reproduced or transmitted in any form or by any means, electronic or mechanical, including photocopying, recording, or by any information storage and retrieval system, without permission in writing by the author. The only exception is by a reviewer, who may quote short excerpts in a review.

No affiliation with, or endorsed by anyone associated, or in any way connected with any amusement park, company, or product listed herein.

Although the author has attempted to exhaustively research all sources to ensure the accuracy and completeness of information on the subject matter, the author assumes no responsibility for errors, inaccuracies, omissions, or any other inconsistencies herein. We recognize that some words, model names and designations, for example, mentioned herein are the property of the trademark owner. We use them for identification purposes only. This is not an official publication.

The purpose of this text is to complement and supplement other texts and resources. You are urged to read all the available literature, learn as much as you can about the field of engineering, and adapt the information to your needs. There may be mistakes within this manual. The information contained herein is intended to be of general interest to you and is provided "as is", and it does not address the circumstances of any individual or entity. Nothing herein constitutes professional advice, nor does it constitute a comprehensive or complete statement of the issues discussed thereto.

Therefore, the text should only be used as a general and introductory guide and not as the sole source for roller coaster engineering.

Readers should also be aware that internet websites listed in this work may have changed or disappeared between when this work was written and when it is read. Side effects of reading this book may include dizziness, nausea, or the sudden urge to visit a theme park.

Figure 1 – Phoenix at Knoebels

Coasters 101: An Engineer's Guide to Roller Coaster Design is dedicated to all aspiring roller coaster designers. Never give up on your dreams!

"You have to be a little bit mean. Sometimes you have to be a little bit sneaky. You get them going on a nice straight track and they think 'This looks smooth,' and then you dip it down a little to give them a good jolt. Or you have it so that when they go over a hill it looks like they're going to get their heads chopped off at the bottom." -William L. "Bill" Cobb, coaster designer

Chapter 1: Coasters 101

Have you ever wondered what it takes to design a roller coaster? It's not as easy as computer games make it out to be! A roller coaster, while simple in terms of the sheer thrill it gives to those who ride it, is a complex three-dimensional puzzle consisting of thousands of individual components.

There is no apprenticeship or degree specifically targeted at engineering a roller coaster. Whether it's wood or steel construction, the job is an interdisciplinary mix of structural, mechanical, civil, and electrical engineering. No two roller coasters are designed the same way. Each project has its own unique set of circumstances and challenges. A perfect blend of engineering and art, roller coaster designers spend countless hours creating and tweaking ride paths to push the envelope of exhilaration all while maintaining the highest standards for safety.

The academic definition of the term *101* means "the introductory material in a course of study." Coincidentally, the term *101* used to be as common at theme parks as it is on a campus. If an attraction shuts down for whatever reason, a string of messages will fly back and forth between the operations and maintenance staff. This used to be accomplished via hand-held radios or walkie-talkies. To avoid a guest overhearing the details of a broken-down ride it was common practice to use the code number *101* rather than announce "The ride is down" to anyone within earshot. Nowadays, with texting and smart phones replacing radios, it's no longer as critical but is still a convenient coincidence.

Coasters 101: An Engineer's Guide to Roller Coaster Design examines the numerous diverse aspects of roller coaster creation including some of the mathematical formulas and engineering concepts that go into designing a multimillion-dollar thrill ride. This introductory text is meant to give a better understanding of the thought process that goes into designing a roller coaster, from concept to creation. Please keep in mind that some equations are generally simplified. Although a basic understanding of physics terms such as velocity, acceleration, and g-force is required (and such terms are defined in the glossary located at the end of this book) you won't need an engineering or advanced physics degree to understand the content of **Coasters 101**. Now hang on tight and enjoy the ride!

Brief History of the Roller Coaster

The ancestor of the modern roller coaster is thought to have originated in Russia in the 15th century as slides constructed of wood and covered in ice. These "Russian Mountains" were built to keep the local residents entertained during the long Russian winters. Over time, the ices slides became more and more sophisticated and grew to be as tall as seventy feet. Sand was used at the end of the ramps to slow the sleds down. Originally, these early thrill rides could only be experienced during the winter. At the start of the 1800s, wheels were added to the sleds enabling year-round rides. These roller coasters in Russian are called "American Mountains."

The next major advancement in the evolution of the roller coaster came in the early 1800s in France, when carts with wheels were first used on grooved tracks preventing the carts from flying off. Next, primitive cable systems were introduced to pull the cars to the top of the hills. The Aerial Walk (*Promenades Aeriennes*) was the first full circuit roller coaster, meaning the track was an uninterrupted closed loop.

Scenic railways appeared in America in the 1870s as a method for railroad companies to increase their business. Even more improvements were made such as automatically being lifted to the top of the tallest hill, wheels on the underside to lock the cars to the track, and two side-by-side rides that raced one another.

Coasters 101: An Engineer's Guide to Roller Coaster Design

Figure 2 - *Promenades Aeriennes, Paris 1817*

The term "roller coaster" is thought to have originated from American thrill rides in the 1880s that used hundreds of rollers mounted to slides or ramps on which a sled or toboggan rode. Passengers coasted along on rollers, hence the name *roller coaster*. While providing a name to these creations, this particular ride had little else to do with the physical development of the roller coaster.

In 1912, the first coaster with upstop wheels was introduced by John Miller. This design held the coaster train on the track and allowed for more speed and steeper hills. The first use of lap bar restraints was in 1907 on the Drop-the-Dips at Coney Island in Brooklyn, New York. In 1925, the Revere Beach Cyclone became the first roller coaster to break the one-hundred-foot-tall barrier.

The 1920s were the "Golden Age" of roller coasters in America. An estimated 2,000 wooden coasters were built before the Great Depression reversed the process. The amusement industry was hit hard during the thirties and roller coasters and amusement parks didn't bounce back until the 1970s. One of the few bright spots between 1940 and 1970 was the birth of the Six Flags amusement park chain in 1961, the future home to many innovative and groundbreaking roller coasters. In 1972, The Racer at Kings Island in Mason, Ohio sparked a renewed interest in amusement parks and ushered in the "Second Golden Age" of wooden roller coasters.

The first tubular steel coaster, the Matterhorn at Disneyland, opened in 1959 (previous rides made of steel did not use tubular track). The Matterhorn also pioneered control systems by being the first coaster to safely allow multiple trains to operate on the same track at the same time due to the use of individual brake zones (called blocks). Another amazing fact is, despite being the first of its kind, the ride was designed and built completely in less than a year – an incredible feat considering paper rather than computers was used for the design process!

The Matterhorn also helped pave the way for looping roller coasters. The French, among others, had experimented with looping coasters in 1848 but the circular loops they designed and constructed put too much strain on the riders (as much as 12gs!). The inversion wasn't perfected until the 1970s. The Corkscrew at Knott's Berry Farm became the first modern looping roller coaster. Many others soon followed. The strength and versatility of steel track and support systems lead to more innovative designs and seating configurations. The 1990s saw a massive surge in steel roller coaster creation leading historians to call it the "Golden Age of Steel" coasters. Today, most roller coasters are classified as "steel" rather than "wooden," though as you'll read later the line between the two types has become very blurred.

Modern day roller coasters are very precisely designed. Long gone are the days of trial and error. From paper and pencil, to bending coat hangers, to realistic 3D modeling, the method for designing the ultimate scream machine has evolved enormously. The result, however, has stayed the same.

See the appendix for more notable dates in roller coaster history.

Coaster Classification

Since Disneyland's Matterhorn opened in 1959, roller coasters have been divided into two main categories: steel roller coasters and wooden roller coasters. According to the Roller Coaster Database, there are over 5,500 roller coasters operating worldwide today. Of these, only 184 (or less than four percent) are classified as "wood" coasters. Over the years, the line between the two has been blurred. But with all the latest technologies and innovations, maybe it's time to redefine the different types of roller coaster.

The traditional defining factor for roller coaster types used to be solely based on the material that the rails are constructed from and not what the supports are made of. From *The American Roller Coaster* by Scott Rutherford (2000): *"To begin, there are two basic types of roller coasters: the classic wood-track rides and those sporting track fashioned of steel.... the track construction itself – not the supporting structure – defines the category into which a coaster is placed. If the track is made of laminated wood on which steel strap rails are mounted, it's considered a wood coaster. If the track is made entirely of steel components, it's a steel coaster."*

Generally, steel roller coasters are defined as a roller coaster with track consisting of tubular steel rails while wooden coaster tracks are made from layers of laminated wood. There are pros and cons for building either type.

A steel roller coaster is usually completely fabricated from steel, including the rails and supports, and the total material weight could be over a thousand tons. Tubular steel rails are formed by heating and then permanently bending steel pipe into the desired shape by running the straight pieces of steel through a series of rollers that mold the pipes into shape. The difficulty is metal bends at either its weakest point or where the strongest force is applied over a span. For a smooth ride the rails need to be extremely precise, within a tenth of millimeter of their designed shape. For this reason, track manufacturers are very secretive about their exact steel bending processes.

However, the bending process can cause significant fatigue in the material of the pipe. The roller coaster track needs to support static loads during construction and installation and dynamic loads as a coaster train travels along it. During the lifetime of the roller coaster, the stresses due to the aforementioned loads along with the initial manufacturing stress results in the pipe needing to be eventually replaced — a very costly endeavor. Many older steel coasters have been closed, torn down, sold for scrap, or closed for a lengthy refurbishment during which the entire original track is completely replaced. As a result, some steel roller coaster manufacturers are going away from bending pipes. Instead, they cut out flat pieces of steel, bend them into shape, and then weld them together to form a box or "I" shape.

Technically speaking, every wooden roller coaster is a "steel" coaster because all the wheels ride on bands of steel. This track steel sits on top of a stack of eight pieces of wood, "the stack" being what defines it as a true wooden coaster. The top two pieces of wood are wider than the boards they sit on so the safety or upstop wheels can run below them thus preventing the vehicles from leaving the track. To save on costs, the "track steel" is used on the underside of the boards only where it is absolutely needed. Locations such as valleys and dips are left bare of steel because the upstop wheels have no chance of ever touching the track. To keep the cars from wedging in the turns, there is normally a small overall clearance permitting around 3/16 to ¼ inches of potential movement between the guide wheels and the track steel mounted to the insides of the top two boards.

Figure 3 – A typical wood coaster "stack"

Traditional wooden coaster track is assembled on the construction site by carpenters who are exposed to the elements and must deal with weather while trying to complete the arduous task of assembling the coaster. Laminated wooden planks are stacked on top of each other and steel running rails are bolted to the top layer. Southern yellow pine is the wood of choice for most wooden roller coaster designs because it has a good strength-to-ductility ratio (more on materials later). A tool called a "track gauge" is used to ensure that the track rails are at the exact distance from each other. Keeping proper spacing ensures the train doesn't hunt, or slam back and forth, so the passengers have a smooth ride, and the maintenance department has less wear and tear on trains and structure to repair.

A wood coaster's support structure is primarily constructed of four by eight-inch wood posts that are bolted together in cross sections known as "bents." Horizontal and diagonal braces are bolted to the posts to form the bents, which are then erected and bolted to each other using members known as "ribbons" that join the bents together using steel plates. Older wood coasters had smaller members and were mainly held together with literally tons of nails. Modern rides are bolted together, providing a better connection with higher strength for the increasingly more dynamic rides.

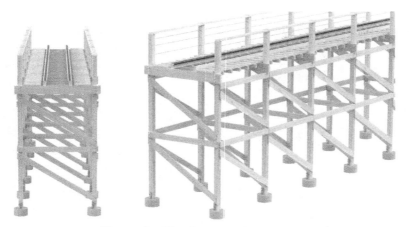

Figure 4 – Wood coaster bent construction

There's a saying about wooden roller coasters: "If it doesn't shake it's going to break." Wooden roller coaster structures are designed to sway a couple of inches as the train goes racing by, especially in tight corners and high g-force locations. Think of it like this — when you jump off a tall object you land safely by allowing your legs to flex and bend at the knee. Otherwise, if you kept your legs straight, you might shatter your leg bone or bust your knee joint. This same basic principle applies to a wooden roller coaster. The structure must be allowed to give and flex like a shock absorber to keep it from internally shaking itself to pieces. It's a delicate balance: the structure needs to move and flex some but not too much. Overtime, if the structure moves too much, the ride centerline could move a few inches then riders will feel it, and not in a good way.

Another expression is an amusement park "never stops building a wood coaster." The wood directly supporting the track must be replaced every four to seven years. Wooden coasters have relatively large tolerances and can sometimes deviate from the desired design over time due to the manual fabrication and maintenance processes. These deviations and large tolerance combined with poorly tracking wooden coaster trains often result in rough and bumpy rides. Wooden roller coasters annually require large amounts of time and money to keep the ride in top operating condition through regular re-tracking, track lubrication, and maintenance. Wood can be a very versatile material to use but it comes with many shortcomings.

Wood Versus Steel

The endless debate among coaster fans is which is better: wood or steel? While steel coasters can give a glass smooth ride, wooden coasters are wilder and provide that out of control, "Oh my gosh this thing's going to break apart right under me" feeling. Wood is character; steel is precision. Steel roller coasters are designed very precisely with tight tolerances, resulting in a more controlled ride experience. They can maneuver high g-force elements complex inversions and tend to be taller and faster. These advantages come with a price though, as the initial cost for a steel coaster is typically greater than that of a wooden coaster. However, the wood coaster may end up costing more in the long run due to more intensive maintenance requirements.

Figure 5 – Steel Behemoth versus wooden Minebuster

Today, a large steel coaster averages an initial cost of $20 million USD, while a wooden coaster will only set a park back just a cool $10 million (theming not included!). To keep costs down, it is desirable to reuse as many components as possible from one roller coaster to the next. With a steel roller coaster, every single support is unique, and custom designed from scratch for that single ride. Thus, it's more affordable for that manufacturing company to be able to sell the exact same steel roller coaster to multiple amusement parks. Wooden coasters designed by the same company share a similar support structure. Unique wood coasters can be built by reusing standard pieces of wood and connectors repeatedly. It's like building a ride with a set of Legos - all the pieces are the same, but the end result is different. The more components that can be standardized, the more affordable the coaster can be. Less design time too.

One reason for the big disparity between the number of wood and steel coasters in existence is the perception that it's harder to market a wood coaster. What's the first word that pops into your head when someone says the words "wood coaster"? For most people, the answer is "rough." A wooden coaster needs a unique aspect the marketing department can use to sell the ride as being special and not like anything else. You really need a good marketing hook. Hence the reason inversions are being added to wood coasters. But there's more than one way to create a unique ride. With limited space and budget, ZDT's Amusement Park in Texas opened the world's first wooden shuttle coaster where the track twists upward into a nearly vertical spike then sends the cars back through the course in reverse.

Why are wooden roller coasters rougher than steel coasters? For several reasons but it starts with the design. Wood coasters designed prior to computers used rudimentary design techniques resulting in painful transitions. Designers often didn't consider impact loads when going from a straight section into a curve. Then there's the nature of wood itself which is not consistent, you have grain inconsistencies, and some parts are less dense. Then there's the vehicles, which we'll get into more later. Most older wood coasters use PTC trains which are giant rectangles. They're solid cars but unforgiving around curves, beating up the track instead of steering smoothly through them. They work well on out and back coasters but not on twisters.

Recently, the line between steel and wood coasters has been blurring, especially with the advent of prefabricated wooden tracks. Advances in wood coaster design and construction allow wooden coasters to perform more like steel coaster with more intense maneuvers and even inversions but at a lower initial cost. Wooden coasters have more steel components than ever before. Particular coaster manufacturers, like Intamin, now industrially prefabricate the wooden track segments which can then be shipped to the construction site and are mounted on the support structures, similar to a modern steel coaster. The rails can be milled to a precise form with a very tight tolerance in a machine shop, not only improving the ride experience but also reducing construction time as well as service life. The pieces of track are then fixed to each bent, creating a more rigid system as opposed to the traditional wood coaster track which floats more freely on the structure. In fact, the ride becomes so smooth that enthusiasts argue that these are no longer wooden roller coasters and should be classified as something entirely different.

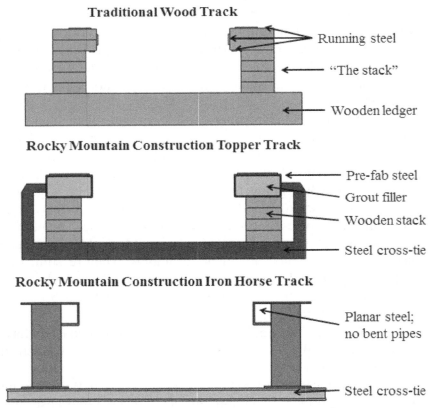

Figure 6 – Different types of track construction techniques

Rocky Mountain Construction (RMC) has developed a new technology called "topper track" where the top two pieces of a wood coaster's stack are replaced with steel and filled with a high strength concrete grout. This new steel topper is very precisely manufactured in a factory before being shipped to the construction site where it is bolted to the support structure. This type of coaster is still labeled as wooden due to the use of a wooden stack, though not everyone is convinced it shouldn't be given its own type of classification such as "hybrid." Adding more confusion to the situation is RMC's Iron Horse steel track that can be built on an all-wood, all-steel, or mixed support structure. In fact, this track has mostly been used in converting old, tired, traditional wooden coasters into new, sleek, steel machines. Roller coasters like Dollywood's Lightning Rod contain both wooden Topper Track and steel I-box track making it nearly impossible to classify as wood or steel and landing squarely in the hybrid category.

Inversions on wooden coasters are becoming more common. Outlaw Run at Silver Dollar City, the first "from-the-ground-up" topper track coaster by Rocky Mountain Construction, performs three inversions – a wooden coaster first since Son of Beast's steel loop was removed in 2006. Also in 2013, Hades at Mount Olympics was renovated by the Gravity Group to add a corkscrew inversion. The stack on Hades is all wood but the support structure for the entire ride is completely made from steel.

Figure 7 – Hades added an inversion to become Hades 360

Another new type of track is the recent collaboration between Skyline Attractions and Great Coasters International on the revolutionary new, weld-free Titan Track. Titan Track is steel roller coaster track that could be used for re-tracking high stress areas on pre-existing rides, adding unique new elements to older attractions, or even building new ground-up rides. In August 2022, Worlds of Fun amusement park in Kansas City announced Zambezi Zinger, a new galvanized steel and wood hybrid coaster. To market a mostly wood coaster, the park decided to go with a unique spiral lift hill and tapped into nostalgia by paying homage to a former steel coaster of the same name. Even new wood coasters are gradually becoming more like steel coasters.

Design Software and Computer Technology

Modern roller coasters are designed using the latest in computer technology. What used to be done with paper, pencil, and drawing boards is nowadays done on a computer. The universal tool for the engineer is Computer Aided Design (CAD) software. Changes and iterations are made on a computer in seconds. Programs such as AutoCAD are used to draft, adjust, and detail designs in accordance with ride standards. 3D CAD software, such as CATIA, SolidWorks, or AutoDesk Inventor, allow designers to have every step in the roller coaster design process contained within one computer program, including:

- ❖ 3D modeling
- ❖ 2D manufacturing drawings
- ❖ Kinematic simulations
- ❖ Finite element stress analysis (FEA)
- ❖ Seismic analysis of fixed or isolated structures
- ❖ High-resolution image renderings for marketing and sales purposes

Some companies rely on Model Based Definition (MBD), an environmentally friendly practice that saves paper by using fewer 2D drawings. All the required manufacturing information is contained within the 3D model as annotation data. Other companies employ proprietary software that optimizes the track layout using elaborate numerical algorithms to help keep the g-forces on the passengers below required safety and customer defined limits. These programs can analyze the dynamic and static calculations of the track and automatically determine where supports should be placed. Standard parts like bolts, springs, nuts, screws, and washers can be taken from standard part libraries or catalogs and bills of material can be derived directly from the model and inserted into a drawing.

Contemporary coasters are built to the highest standards, quality, and tolerances. Modern 3D CAD systems can help the engineers design the ride within the limits of biodynamic tolerances of the passengers. A simulation of the coaster, based on the CAD data, can illustrate the g-forces and the dynamic behavior of a ride long before a prototype has been built. The roller coaster is virtually assembled early in the design phase to test functional relations and mechanisms, the ride's clearance envelope (the area within possible reach of the passengers in the vehicle), as well as collision detection of components. Engineers can change the shape of the track and see how it changes the forces on the riders in real time as well as the total cost of the ride.

Utilizing CAD software is an enormous improvement over methods that were employed even thirty years ago. The time required to create a roller coaster from scratch would be quite lengthy without using a computer because of the enormous number of calculations required. Changes and iterations can be made very rapidly using the computer. Multiple track variations for a ride may be presented to the customer to choose their favorite layout. Computers can give detailed information about every point on the ride. Banked turns are formed with smooth-curve algorithms while transitions from straight to curved track are performed with splines – a process that is made easy when the computer handles the calculations.

After the CAD work is complete the production data is electronically sent to the manufacturer or fabricator. Even with this high-powered CAD technology, the transition from promising mockup to actual scream machine is never accomplished without a few surprises. There are thousands of variables in play that can wreak havoc on the best of intentions and it's impossible to adequately address them all solely during the imagineering phase. These unknowns can delay a project significantly and should be budgeted for in the project's master schedule.

Coaster components must be designed with the manufacturing and installation processes in mind. Engineers generally design parts with a tolerance range. If the manufactured part's properties fall within the specified range, the function of the component shouldn't be affected. Tolerances are specified to allow for variability and imperfections within the manufacturing process. It can be extremely difficult to manufacture parts exactly as they are designed. Fusing two pieces of metal together can be very challenging, especially when they must be at a very specific angle for them to fit together. Though 3D CAD software is a powerful tool, designers must be careful and thoroughly think through all aspects of their design so as not to fall into one of the numerous pitfalls when relying on computer software.

For example, consider designing a steel support column and track segment that are going to bolt together with eight, half-inch diameter galvanized steel bolts. In the 3D CAD model, the bolt holes on the support will be perfectly aligned with the holes on the track one hundred percent of the time. In reality, when that support column is manufactured the holes are potentially not going to be exactly where the designer modeled them. They could be off-center by a fraction of a millimeter or more, the hole could end up being not perfectly round, or the angle of the hole might not be normal to the mating surface. Engineers must account for this in the design by making one of the sets of holes on the track or support slightly larger than the diameter of the hole on the other mating component. This way, if the holes were not drilled exactly to specification, and the holes are slightly off center to each other, the bolt can still be inserted through both holes.

Another scenario: A real piece of manufactured track may be 0.02 inches longer than it was designed on the computer. While this may seem insignificant, over the course of a near mile long layout the inconsistencies add up, making it impossible to secure the last track segment to the first, and a contractor quickly has a substantial problem on his hands. This dilemma is known as "tolerance stack-up." Engineers must plan for the best- and worst-case scenarios by studying the dimensional relationships within an assembly. Generally, the more precise the tolerance, the harder it is to achieve, thus the higher the cost to maintain that quality.

What software do roller coaster engineers use?

Roller coaster designers use a combination of commercial software and in-house programs. The commercial software is mostly the same platforms used at any other non-coaster engineering firms. Engineers don't need to know how to use every single CAD system. Chances are, if you can learn how to use one the others should be easy to pick up if needed. The following table includes commonly used software in the amusement industry. Aspiring coaster designers should become familiar with at least one system in each category.

Real roller coaster designers do use NoLimits Coaster Simulator, usually not for engineering purposes, but to quickly and easily create realistic ride proposals to send to potential customers for bidding on a project. NoLimits 2 utilizes a node and spline method to create roller coaster tracks. The nodes are adjusted individually then the software calculates a smooth Bézier (parametric) curve to create the final track shape. The location and roll of the spline at given intervals can be imported and exported into other programs like Microsoft Excel, which can then be imported into 3D CAD systems like SolidWorks.

Another tool often used in conjunction with NoLimits is FVD++ (Force Vector Design) which allows the user to create coaster layouts using mathematical formulas. The shape of the elements is generated by calculating the desired forces on the riders rather than the designer specifying the track radii resulting in a layout consisting of constantly changing radii.

Ultimately, the methods for generating the track layout on the computer and how it is manufactured are what sets the different manufacturers apart. Different companies may use similar track styles or support structures, but the software and manufacturing abilities have the biggest impact on the feel of the ride experience. Therefore, both are usually closely guarded trade secrets.

Type	Primary Recommended	Secondary Systems
2D	AutoCAD	DraftSight
3D	SolidWorks	AutoDesk Inventor CATIA V5 ProEngineer Solid Edge
Rendering	AutoDesk 3DS Max	Rhino 3D
Math Computations	Maple	Mathcad
Programming Languages	VBA	Matlab AutoLisp C#
Structure Analysis	RISA 3D	ANSYS
Concept/ Marketing	NoLimits 1 & 2 Coaster Simulation	

Imagine how long it would take to model every single piece of lumber on a wood coaster. For this reason, designers have produced their own in-house programs to auto-draw many of the components needed to virtually assemble the ride, such as all the wooden bents and standard bolts. These programs may be separate pieces of software or code written as macros in existing software packages (a macro is a program that can imitate keystrokes or mouse actions and are used to replace manual repetitive tasks).

If you want to get a step ahead as a roller coaster designer, you could learn how to program macros within whatever modeling software of your choice. Using Visual Basic for Applications (VBA), or any other programming language, allows you to write macros to automate repetitive process and increase your design efficiency. It's not a required skill but it could be very advantageous to your professional career no matter what field you end up in. You should learn all the Microsoft Office software (for any engineering job) but especially Microsoft Excel. This is a great program to begin learning how to write macros because there are numerous free resources on the internet and in the app stores to teach you how to get started.

The steps I would take to learn VBA programming:
1. Learn how to write formulas in Excel spreadsheets – this will teach you basic logic like IF THEN.
2. Learn how to program macros in Excel – relatively easy and many free resources are available.
3. Learn how to program macros in a CAD software – once you learn how to write code for one program it is easier to pick up in another. Eventually you'll be able to have two different software communicate with each other.

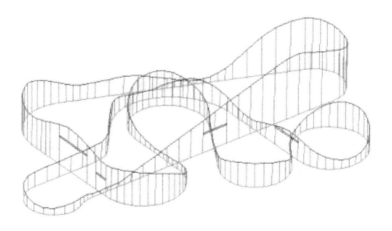

Figure 8 - Rudimentary CAD coaster layout

Chapter 2: Planning

One of the first steps of any project, including coaster design, is determining the project schedule. A roller coaster's track can only be manufactured after all the calculations and analyses are completed. Designers first acknowledge the limitations of the vehicle and decide which elements they wish to negotiate. It often takes a few months before the data for the track can be sent to the fabrication shop or to the structural engineers who design the supports. However, there are instances during coaster construction where the design on one section of the ride is still undergoing modifications even after part of the structure on another section has been assembled on site. Keeping up with the schedule is a challenge because there is so much that has to happen, and each process is dependent on upon another task being completed.

Engineering approaches to project management and scheduling include:

- ❖ A GANTT bar chart is helpful when laying out the varied tasks associated with a given project.
- ❖ PERT (Project Evaluation and Review Technique) is another commonly used planning and scheduling helper. PERT involves analyzing the tasks needed to complete a given project, especially the minimum time needed.
- ❖ In conjunction with Gantt and PERT, engineers may also use CPM (Critical Path Method) which is an algorithm for scheduling a set of activities within a project.

Cost, time, and scope need to be defined at this point. The schedules can be created via any of several software programs, including Microsoft Excel, Microsoft Project, Google Docs, Outlook Calendars, and so on.

Roller coasters are created by a team of people or companies who are each assigned a different task. It depends upon the size of the project but know that roller coasters are designed by as few as four engineers. The team to assemble, test, and market the ride is much larger. The timeline could be between one and five years. To understand what everyone is doing and when they will be done with the different phases of the project, schedules are used and shared by the team. The timeline is constantly being updated and revised due to the many unforeseen issues which so often arise during a project as large as a roller coaster. Communication between team members is crucial to success. Engineering alone may take more than 3,000 hours of work for a single ride. Designing an entirely new coaster concept may take as long as three to four years while a medium-sized wooden may be designed in as few as nine months. Granted, a lot of older roller coasters were designed and built much more quickly than some of the newest machines, but that's simply because safety regulations were nowhere near as strict as they are today.

After months of design and planning work, a roller coaster's construction schedule may follow this general guideline:

Survey: A week or two of surveying the lay of the land and plotting out where the foundations will go.

Foundations: Two to four months of digging and pouring the foundations, after which they must set to cure for a specified period of time.

Structure: Five to seven months of assembling the major structural components of a ride including all the bolt tightening. For a wooden roller coaster, there could be as much as 90,000 bolts and over 4,000 pieces of wood beams.

Tracking: An average sized steel coaster may have between fifty to one hundred segments of track to piece together on the job site. After the structure begins to be completed on a wood coaster, crews of carpenters can begin "tracking" or stacking layers of wood onto the supports and laying the track steel on top.

Electro-Mechanical: Three to six weeks of installing all the mechanical components including the lift motor, chains or LSMs, brakes, station gates, kicker tires, transfer track system, lap bar release system, and all of the electrical work and control systems. Any theme related elements are also installed, like audio-animatronics, on-board audio systems, or static sets.

Test and Adjust: All systems previously mentioned are then extensively tested. The operating crew is trained during this time on how to operate the ride and the maintenance department on how to care for the ride.

Operation/Marketing: Before the public gets to take a spin, the park will usually invite members of the local media out to take a ride, followed by a grand opening ceremony. The marketing department now is responsible for making the local fans and residents aware of the expensive new addition. From a theme park's perspective, the easiest way to market a new coaster is with a "-est" world record. Tallest, fastest, steepest, longest, loopiest, family friendliest. It doesn't even have to be a world record. "Fastest wooden roller coaster in Northern California" will do.

Of course, the construction plan almost never goes exactly according to schedule. Problems spring up, like digging a hole for an underground tunnel and running into an underground spring.

Design Constraints and Considerations

Millions of people visit theme and amusement parks each year. Roller coasters have long been the main attractions on the midways, with the world's coaster count always growing. The experience of controlled falling is scary but exhilarating; the appearance of danger through speed and sensation provides an incredible adrenaline rush. Coaster enthusiasts strap themselves into these scream machines simply because they love being scared in a safe environment. The sensation of weightlessness or an experience of being thrown from the train reinforces the feeling of danger. No, roller coasters are not thought of by an evil genius, designed with the sole purpose of scaring the patrons to death. Roller coaster engineers make the riders feel extreme forces while keeping them safe and secure.

So, what are we waiting for? Let's begin designing the biggest, craziest, uber-fantastic, world-record breaking roller coaster we can imagine!!!

As Lee Corso would say, "Not so fast!"

The first thing any aspiring roller coaster designer must understand is you're not going to design whatever it is you want to. A real-world roller coaster is designed within a strict set of constraints and the designer will have little or no control over most of these boundary conditions. Besides the obvious laws of physics and safety standards, which we'll get into more detail later, here are a few of the major considerations that must be evaluated:

Customer Requirements - Roller coaster designers must design what the customer wants. The amusement park is investing millions of dollars to build the roller coaster and they have a target audience to please as well as attendance and monetary goals to attain. Satisfying the customer is the number one priority; otherwise, the company won't be designing or manufacturing any other coasters. The design must provide the ride experience desired by the client, including the basic layout and all the defining elements. This is called the "design intent."

Questions need to be answered by the client: who is the target audience? Is this a family ride or a high g-force thriller? What type of vehicle or seating configuration will be used? What is the desired Theoretical Hourly Ride Capacity (THRC)? Will there be any theme or show related elements? What are the track switch requirements, maintenance bay storage, and preliminary facility interfaces? How will it reach its maximum potential energy?

Budget Allowance - There is no Sandbox mode in real life like there is in a video game. The client wants to make a return on their investment. Coaster enthusiasts often forget that theme parks are a business. Offering fun and entertainment is how they make their money, but the bottom line must still be satisfied. Roller coasters not only initially cost millions of dollars to build but they will also be a drain on the maintenance department's budget for years to come as wheels and other components will constantly need to be replaced. How many operators will it take to run the ride? What kind of training will they need? Above all else, parks are looking to make a return on their investment (ROI).

Location Restrictions - When a theme park decides it's time to build a new roller coaster one of the first decisions is where to place the ride. Space available, height limitations, terrain, and existing structures to avoid are all considered. What does the terrain look like? Will the station be located at the highest or lowest point on the ground? Certain rides must fit into a given volume where other rides are literally built on top of a flat piece of land with no restrictions (like a former parking lot). Flat, open sites have almost no constraints and allow a completely free flow of ideas for layouts and elements. Other sites almost naturally dictate most of the ride's layout and elements due to the existing obstacles. Some amusement parks try preserving trees as much as possible. While admirable, this directive places another layer of complexity to the design and construction of the coaster.

A new ride's location may be chosen to balance out guest traffic in a specific area of the park or to help revitalize an older section with a brand-new marquee attraction. Sometimes the client wants the highest point of the coaster to be in line with a major pathway to act as what Walt Disney called "a weenie," a beckoning hand to draw guests towards it. Cedar Point created the ultimate front gate by twisting the GateKeeper's track upside down over the entrance plaza.

Roller coaster structures are attached to the ground, making location critical to design. This also makes them highly susceptible to standing water and associated corrosion, as well as storms and natural disasters. A site visit (or several site visits) is the best way to get a feel for the project at hand. Walking around the site and talking with the client gives the designers a solid idea of what terrain they must work with and what the park really wants. Often the site with the most challenges results in the best rides.

City Limits - Even if a theme park wants to build a colossal coaster the local city planners may not allow it. Countless incredible coaster concepts have been denied because of noise level concerns from the residents living next to the park.

Several theme parks have unique challenges to overcome. Alton Towers in the United Kingdom is not allowed to build any rides taller than the tree line. Initially, this restriction might seem like it must result in some lame coaster designs when just the opposite is true. It forces the designers to be more creative. Giant pits and trenches were dug deep into the earth for Nemesis, Oblivion, Air, and The Smiler and, by no coincidence, they are some of the highest rated coasters on the planet. To take advantage of all available land Waldameer park in Erie, Pennsylvania built a portion of the Ravine Flyer II wood coaster over a four-lane highway using a double arched bridge. In consideration of a nearby residential area, Mystic Timber's far turnaround was relocated during design, and a tunnel was added to further muffle sound disturbances for Kings Island.

Selecting a Manufacturer

When amusement parks are planning a new mega coaster, they may ask several different roller coaster manufacturers for their ideas. The designers will submit preliminary plans for the coaster including layout concepts and cost. Roller coasters are a large investment and parks must do their research and due diligence before selecting a supplier. Often the maintenance and engineering teams from the park will travel to other amusement parks around the world to test out a similar ride before signing a contract. When the park is mulling over which manufacturer to select, they must reflect on the following points:

Number of years in business: Does this company have years of experience? Do they have a proven track record? Have they continued to innovate and improve over the years?

Cost: How competitive are their prices?

Customer Service: Will the manufacturer be able to supply the client with customer service for the duration of the coaster's useful life? Do they have a reputation of doing so?

Safety: Does the manufacturer participate in safety committees such as ASTM? Have other installations had any major accidents?

Quality: Are previous customers satisfied with the quality of their roller coaster? Do they have Quality Assurance procedures documented?

Manufacturing Capability: Does the designer manufacture the rides themselves or are they a middleman that outsources to other companies? If the supplier goes out of business will the park still be able to get replacement parts for them?

Availability: Is the desired manufacturer available and able to meet the park's timeline? Fabrication time at the factory may need to be reserved months or even years in advance. A story told in Evan Ponstingle's *Kings Island: A Ride Through Time* explains when Cedar Fair decided not to build a B&M hyper coaster at California's Great America, they didn't want to hurt their relationship with B&M so they didn't outright cancel the order. Clermont Steel Fabricators had already reserved shop time to build them a coaster, so they simply changed the location and the project morphed into Orion at Kings Island.

While they may use the same CAD software, one difference between an engineering firm and a coaster designer or manufacturer is the lack of a proper sales or marketing department. As Rocky Mountain Construction have stated "We love that we don't have a sales team selling our rides, we are very lucky to have a product and a reputation that sells itself...so we don't seek our future clients. The clients come to us." At this point RMC is well-known and has a very in-demand product. The same cannot be said for other suppliers that might need to put more effort into convincing potential customers to take a chance on them.

Primary Product	Designer / Manufacturer	
	North America	Europe
Steel Coasters	Chance Rides Premier Rides S&S Worldwide Rocky Mountain Construction	Bolliger & Mabillard Gerstlauer Intamin AG Ing.-Büro Stengel Mack Rides Maurer Söhne Schwarzkopf Vekoma Zamperla Zierer
Wood Coasters	Gravity Group Great Coasters International Philadelphia Toboggan Coasters Rocky Mountain Construction	

An amusement park looking to build a new scream machine will often accepts bids from several coaster design firms and pick the one with the most unique design or best price. A coaster design firm may bid on more than one hundred projects per year. Of those, maybe twenty of the proposals are taken seriously and move ahead to the next step. Finally, maybe only four or five of those projects are greenlit and become a reality.

Another key thing to understand is everyone does business differently. Not all manufacturers do things the same way and not every park or family entertainment center purchasing a roller coaster follow the exact same process or procedure. Some will be more detailed and standardized than others. Each new customer will be a learning experience.

Layout and Profile

A potential roller coaster buyer will approach the designer in a variety of ways depending on their individual situation. They might come with a set budget and the designer will provide a custom ride within that budget. The park could also approach the designer and ask for a new coaster or overhaul with certain record-breaking elements incorporated into the design. The designer can provide a few different variations with a cost for each one, giving the customer a variety of options. Or a park could come with very few limitations or parameters and the designer creates a few variations of a new coaster for them to choose from.

Once the approximate location of the coaster has been chosen a preliminary pathway is created. This centerline is drawn onto an overhead image of the site to get a feel for the general layout of the ride. The centerline may change later in the process, but for now serves as a starting point. Often, the centerline path is physically marked on the site with spray paint or flags. Balloons raised to specific heights help the park and the designers visualize the coaster before it is built. The station is rarely at ground level due to required maintenance access and room for equipment below.

When the general path of the ride is known, all the ups, downs, and loops are added into the mix. This is known as the profile and g-forces are now taken into consideration. The profile is designed to work with the centerline to deliver the right forces to the riders. During this phase the designers decide where the track will be higher and hence slower versus where it will be lower and faster. The combination of the centerline and the profile make up a layout of the roller coaster. Using 3D CAD software, you can turn two 2D sketches (the top view and the side profile) into a 3D dimensional roller coaster model. More than one possible layout may be presented to the customer.

Ride Analysis

There are a multitude of roles for roller coaster designers, including everything from train assembly to CAD drawings, from being covered in grease to being buried in paperwork. It's simply not enough for an engineer to design a ride on a computer then hand it off to the park. There are 2D assembly drawings for construction that must be made, manuals on how to safely operate the ride, more manuals for how to care and maintain the roller coaster for the duration of its operating life, and so on.

The engineers must perform a complete Ride Analysis (RA) of the coaster before it can be approved for construction. This includes:

- ❖ Patron Restraint and Containment Analysis
- ❖ Patron Clearance Envelope Analysis
- ❖ Failure Analysis
- ❖ Design and Calculations
- ❖ Drawings and Records
- ❖ Regulatory Body Reviews

Designers essentially have a giant checklist of conditions that must be verified before the design is approved for manufacturing. This entails several different tests analyzing specific criteria such as patron restraint and containment, clearance envelope, risk assessment, and safety systems. Accepted engineering practices to demonstrate proper system response to component or device failure may include a Fault Tree Analysis (FTA) or a Failure Mode and Effects Analysis (FEMA). The designers must mitigate every potential hazard as much as possible. Every scenario needs to be investigated, including any reasonable foreseeable misusages of the ride by the operators or patrons. A ballistic envelope must be examined to determine if an object falls out of a rider's pocket or is thrown from the vehicle, where it could potentially land. Once all tests and analyzes of the computer models are complete the project can move ahead to the construction phase.

After all the questions are answered and customer requests understood, it's up to the engineers to make it happen. All the mechanical and electrical components of the design must fit together in the real, physical world. The train must make it from station to station with no pit-stops, shortcuts, or derailments. Not only that, but the structure has to support the natural forces that are incurred at every inch of track. This is the manifestation of all the nitty-gritty engineering work and analysis.

Designs that look fantastic on paper don't always translate well to real manufactured parts. Two pieces can't occupy the same space at the same time. Changing a component to satisfy one condition can easily spoil another, and it takes constant iteration to reach a point where all design requirements are met. Sacrifice and compromise are part of the process. Making sure that all manufacturing and safety requirements are satisfied simultaneously, in the same design, is one of the most challenging aspects of roller coaster engineering (or any engineering project, really).

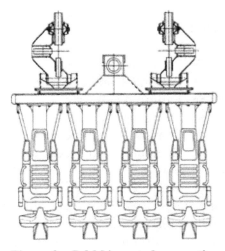

Figure 9 – B&M inverted car section

New and experimental amusement rides should be subjected to a thorough on-paper design process of dynamic (sometimes called "geometric") and structural engineering. Dynamic engineering involves calculating the physics a passenger will experience along the ride path, such as speed, weight, distance, velocity, momentum, gravitational force, centripetal force, friction and so on. Although time consuming, properly performed dynamic engineering yields a safer ride. Once the physics of the design is calculated, the designers adjust the blueprint before the structural engineering begins. Structural engineering is then used to design the foundation and pillars to support the ride.

Chapter 3: Design Principles

The main objective of a roller coaster engineer is to successfully manage all the resultant forces on the vehicle and its passengers. A good design will manage the variables such that most of the energy of pulling or launching the vehicle to the top of the tallest point is dissipated by the time the vehicle returns to the starting position. This way, riders are having fun from the moment they blast off, to when they come in for a smooth, safe return to the loading area. This is all accomplished by effectively managing the drag and other resistant forces.

In this regard, auto racers and airplane pilots have influenced roller coaster design. For instance, one of the most important factors in racing is the ability to minimize the lateral loads on the race car by taking the best (read: smoothest) path through a high-speed turn. This is done by maximizing the radius through the curve.

Piloting technique is much the same. A commercial airline pilot will bank the airplane smoothly and coordinate his turns by applying the proper amount of rudder; otherwise, he would be putting undue stresses on the equipment and causing discomfort to his passengers. Roller coaster engineers have the very same goal. There are numerous similarities in a stunt pilot's tricky aerial maneuvers and coaster inversions. In fact, many names of coaster inversion elements, such as the Immelman loop, come from the aerospace equivalent.

G Forces

Roller coaster enthusiasts may tell you they enjoy the sensation of speed on a roller coaster. In reality, humans can't feel speed; what we do sense is acceleration. When was the last time you felt the Earth's rotation as it spins rapidly under your feet? Flying at 300 mph in a passenger jet feels the same as cruising down the street at thirty in a Toyota Corolla. In other words, without a point of reference it is impossible to tell how fast you're moving. It's not the consistency of any given rate of travel that creates a noticeable sensation, but rather the variance; the stops and starts; the slow-downs and speed-ups.

It's for this reason that a roller coaster designer's goal to create a thrilling coaster should always be to produce as many safe accelerations for the rider as possible. Shake them up; produce a series of highs and lows; press their butts into the seat then pull them out of it; magnify and maximize that feeling coaster junkies love so much, and they'll be standing in line for hours. But how do you design a coaster to achieve the perfect accelerations, that sweet spot where the forces that are fun to feel but not too strong?

G force is expressed as a ratio of the force developed in changing speed or direction relative to the force felt due to the Earth's gravity. G-load refers to acceleration normalized by Earth's gravitational pull (9.81 m/s^2). 1g is the acceleration we feel due to the force of gravity. A 2g force on a 100-pound body causes it to weigh 200 pounds (weight = mg = mass times force of gravity). The term "g-force" is misleading because it refers to acceleration due to gravity. Under Newton's Second Law, F = ma, or force = mass * acceleration. It is used because the weight force is proportional to mass, while acceleration is inversely proportional, so the acceleration of all objects due to gravity is equal.

To keep rider's safe, a standardized table of maximum values for gs have been set by amusement industry safety committees. As the duration increases, the maximum g force limit decreases. For example, 4gs should not be experienced for more than four seconds. Human tolerance of g force depends on the magnitude of the gravitational force, the duration, the direction, and location it acts on the body, as well as the rider's posture. If a roller coaster is speeding through a complex maneuver, multiaxial gs are important to consider as well.

To help keep g forces in check, modern roller coasters are designed not around the center of the rails but the "heartline" of the passengers. As we'll get into more later, when a coaster's track is banked through a turn, the axis of rotation should be near the center of the rider's heart, hence the name "heartline." Now most coaster vehicles or trains have more than one passenger so the heartline may not be in the center of a single rider but rather the average of all the passengers. Moving the rotation axis higher minimizes a lot of unwanted and uncomfortable lateral movement (especially to the rider's heads) creating a more comfortable ride experience.

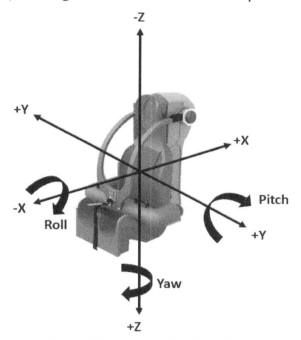

Figure 10 – Acceleration directions

Positive g-forces, meaning those that push your butt into the seat, become uncomfortable for the human body at +5g and may cause the loss of consciousness. Indianapolis 500 racers are subjected to more than 3g's in the corners of their hairpin turns while there are looping coasters that subject passengers to as much as 6g's for a very short period. When humans are under high positive gs, blood is being forced from the head to the feet, eventually causing a grey out, tunnel vision, or black out. On a roller coaster's drop, the smaller the curve radius and the higher the speed, the greater the g-force felt.

Conversely, roller coasters also thrust negative vertical gs on riders causing them to momentarily lift off their seats and become "weightless." As the vehicle flies over the top of a hill the load on the passenger becomes less than Earth's gravity and, in the extreme, could throw an unrestrained passenger out of the car. Scream machines with oodles of so-called "airtime" moments or "butterflies in your stomach" thrills rank among the world's best. The Voyage at Holiday World contains 24.3 seconds of combined airtime over its 6,422-foot-long journey. As a comparison, passengers on NASA's Vomit Comet experience near-zero gs for as much as twenty-five seconds at a time. Negative g-forces cannot be too great because when under high negative g forces blood rushes to the head and can cause "red out."

Airtime is typically categorized further into two categories: floater and ejector. They don't have standardized values but generally floater is sustained airtime between -0.5 and +0.5 vertical gs while ejector is more abrupt, lower than -0.5g but short in duration.

Designers spend countless hours calculating the curves of those airtime hills to produce the best sensation of weightlessness while maintaining safety standards. Most modern airtime hills are parabolic in shape, meaning the radius decreases as it nears the crest of the hill. It's challenging to design within safe accelerations because the train usually doesn't travel at a constant speed over the hill and thus the g forces will vary from front to back, especially on longer trains. The more cars in the train, the greater the distance between the center of gravity and the furthest seats and the bigger the variance in the g forces. This variance needs to be studied when designing the layout because the ride must be safe and not exceed any g force limitations in any seat.

The train moves the same speed but at any given point on the track, the entire train goes at different speeds. The speed over the crest of the hill is higher as the first and last cars traverse it. As the train hits the top of a hill it is going quite slowly. But after it has gone over the top it speeds up. This is because the center of gravity of the entire train is lower, resulting in higher kinetic and less potential energy. As a result, the front and back seats will offer greater spikes in airtime, but the middle of the train will be more consistent (on symmetrically shaped elements).

Figure 11 – A forceful parabolic airtime hill on Storm Chaser

Every coaster fan has a preference of sitting in the front or the back of the train due to the varying ride experiences. There are those who prefer "the sinking feeling" of the first car as it speedily plunges into those sharply angled valleys, and then there are those that prefer the last car for the whiplash thrills of being yanked over the top of the camelback hills.

Besides vertical positive and negative gs, there are also g forces that push riders to either side of the vehicle. Lateral gs are often considered uncomfortable since the seat rarely has adequate or comfortable support for the rider. Banking the track converts laterals into positive vertical loads, which the seat can support more comfortably. Laterals are generally avoided, except in wooden coasters where they add to the feeling of being out-of-control, but in these cases the seats often have extra cushioning and support. The Beast at Kings Island may be the coaster best known for lateral forces, especially with its unique double helix finale.

Kinetic and Potential Energy

How a roller coaster achieves its top speed all comes down to energy. Energy is a conserved quantity that flows when there is a change. Energy is stored in an object: mass ($e=mc^2$) and kinetic energy. Energy may also be stored in a field.

On a coaster with a lift hill, as the train is pulled to the top, energy is being transferred into the gravitational field or stored gravitational energy. Once released, the acceleration due to gravity makes the vehicles coast back to the station. The higher the lift, the greater the amount of potential energy stored in the field. This is shown by the equation for potential energy:

PE = potential energy = energy of position = mass * g * height

PE is potential energy, m is mass (kg), g is the gravitational field intensity, equivalent to the acceleration due to gravity of 9.8 m/s/s, and h is the distance about the ground (in meters).

Because mass and gravity are constant for the train, if the height of the train above the starting position is increased, the gravitational potential energy must also increase. Thus, gravitational potential energy is greatest at the highest point of a roller coaster. The taller the lift hill, the more potential energy is stored in the gravitational field.

As the train accelerates down the hill, potential energy is transferred into kinetic energy, shown by this equation:

KE = kinetic energy = energy of motion = 0.5*mass*velocity^2

There is a difference between velocity and speed. Velocity is a vector that has a direction while speed is just a rate over distance. One thing to note is that if you double the lift hill height, you don't automatically double the max speed at the bottom of the drop. The squared term means that if you double your starting height, the velocity is increased by the square root of 2, which is a factor of only 1.414.

Roller coasters are constantly trading potential energy for kinetic energy. The most important thing to know about energy is that energy is never created or destroyed; it's only transferred. When someone says "energy loss" the energy is not really lost, it's just passed somewhere else. Likewise, saying the kinetic energy was converted to heat is technically incorrect: the energy is not converted, it is transferred.

While the law of the conservation of energy still holds true, a roller coaster is not an isolated system. Some of the energy input into the roller coaster system will leak out to the world through dissipative forces, causing the roller coaster to naturally slow down. Dissipative forces such as friction or drag result in some of the kinetic energy being "lost", meaning transferred to heat or thermal energy or sound.

Total energy = Potential Energy + Kinetic Energy + Thermal Energy

Imagine having three containers, labeled PE, KE, and TE. At the top of the lift hill, the PE container is filled with a liquid, representing energy, while the other two containers are empty. As the train goes down the hill, most of the liquid is poured from the PE container into the KE container, and a small amount into the TE container. To keep the speed as high as possible, you want to keep the KE container full and the TE as empty as possible. The amount of liquid is always the same, it is just poured into different containers.

Roller coasters are all about kinetic energy, the energy of movement, versus potential energy, the energy of position. They use stored mechanical energy rather than an engine to operate. As a vehicle travels downhill it trades its "head" or elevation (think of it as the currency of potential energy) for velocity (the currency of kinetic energy). The maximum speed of a coaster is usually achieved in one of two ways: it is lifted to the top of the highest point of the track by a system and then released or it utilizes a mechanism to shoot or launch the vehicle from a standstill to its maximum velocity. In either method, stored energy is transferred to active energy.

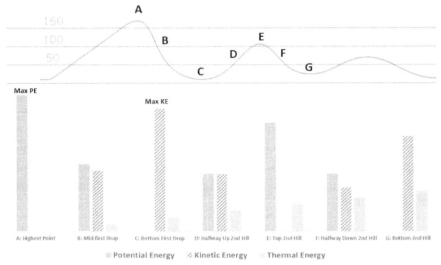

Figure 12 – Relationship between PE, KE, & TE

Friction and Drag

The speed of a roller coaster train depends first on the amount of energy input at the start. This is usually either the height of the lift hill or the launching energy (where velocity, $v = \sqrt{2gh}$). After the roller coaster has reached its highest point, it's all downhill from there. The vehicle's speed is a function of changes in elevation and the various drag factors that tend to slow it down. Why is speed important? If the train is going too slow it may not be able to make it around the entire length of track. If it's going too fast the forces on the riders may be too great and cause injuries.

For a non-idealized roller coaster system, not all of the energy is conserved. Friction is the main cause of energy "leaks" in the system and the reason why mechanical energy is not fully conserved. This is because friction is a non-conservative force. Non-conservative forces are forces that cause a change in total mechanical energy. Friction opposes motion by working in the opposite direction. The friction between the train and the tracks as well as between the train and the air take energy out of the system, slowing the train and creating both heat and sound. This effect is most noticeable at the end of the ride as all remaining kinetic energy is taken out of the system through brakes and converted to heat. Because of the energy leaks due to friction, each successive hill or loop on a roller coaster must be shorter than all the hills or loops preceding it, otherwise the train will not have enough energy to make it all the way around the circuit. If there were no friction, a roller coaster could roll around and around the track forever without ever slowing down.

Listed below are factors that can negatively affect the speed of a roller coaster train and how a roller coaster designer can maximize the kinetic energy.

Aerodynamic Drag

Air resistance, or aerodynamic drag, is a force that acts on a solid object in the opposite direction to the relative motion of that object through the air. Simply and rather obviously, drag slows down a moving vehicle. Drag depends on the square of velocity. When drag is equal to weight, acceleration is zero and the velocity becomes constant (an event known as terminal velocity). The two most common factors that have a direct effect upon the amount of air resistance are the speed of the object and the cross-sectional area of the object. Increased speed or increased cross-sectional area result in increased resistance.

The direction and speed of the wind affects the speed of the train. As you'll see in a minute, wind resistance goes up with velocity squared. If the wind is blowing strongly directly into a car right as it's cresting a hill, it could result in the coaster stalling and rolling backwards, becoming stuck in the low point between hills (called "valleying"). High winds will sometimes force amusement parks to close the ride until conditions are safe to operate. Depending on the direction of the wind, a single car coaster will be more susceptible to wind drag than a train of cars because of the much less weight relative to the frontal surface area. However, in high-speed twisting parts of the track, the drag is increased considerably when the individual vehicles of a train fan out and more frontal area of the train is directly exposed to the wind.

Drag acts perpendicular to a surface therefore the frontal surface areas have a large impact, as does the shape of the vehicle. Numerous scientific experiments have determined that a teardrop shape experiences the least amount of drag. This is due to two factors, both of which should be taken into consideration when designing the shape of any vehicle:
1. The frontal surface area is curved, which allows the wind to transition smoothly from its straight path to around the object.
2. The rear converges at a point.

Now, we probably won't be able to design our coaster car to converge at a point, but we can make the shape as aerodynamic as possible. The losses due to aerodynamic drag could be lessened with an aerodynamically shaped car.

To compute the aerodynamic drag force, we need to know the frontal surface area *A* of the car plus passengers measured in meters squared *(m²)*, velocity *v* measured in meters per second *(m/s)*, ρ (the Greek letter rho) is the density of air (at sea level it is *1.225 kg/m³)*, and *CD* is the coefficient of drag:

$$F_D = \frac{1}{2}\rho v^2 C_D A$$

The drag constant "CD" is determined by experiment as there is no theoretical value for "CD". The type of paint, exact vehicle shape, and all kinds of other variables will affect this number. If an engineer wanted to include air resistance, he would calculate the motion without air resistance for a similar roller coaster, make actual measurements for that similar coaster, and then compute an average.

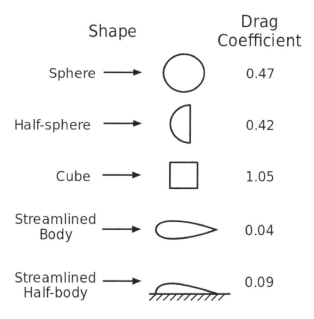

Figure 13 – Measured drag coefficients

However, because the front of many roller coasters is aerodynamically designed (not just a flat plate) and have significant amounts of mass (several thousand pounds), then air resistance will not have a huge effect. Neglecting air resistance is a safe assumption on some roller coasters because the vehicle is so massive relative to the front cross section of the vehicle – the surface area that would contribute most to air resistance.

Of course, this becomes less true for roller coasters with large frontal surface areas. The average roller coaster sits two or four riders across a row with one or two rows per vehicle. The widest roller coaster vehicles seat ten riders across in three rows. Wider trains are commonly found on wing riders and dive coasters. Sitting riders in a row of ten across allows for more people to sit in and experience the ride from the front row. Such coasters often have vertical, ninety degree drops and will even hold the train at the top of the terrifying descent for a few spine-tingling moments (called a "dive" coaster).

Another characteristic of coasters with wide vehicles is the track may not rotate around the heartline through inversions but rather will rotate around the center of the track instead. Inversions on these types of rides are rather drawn out and taken at a slower rate of speed to compensate and hold the forces on the riders at an acceptable level.

Figure 14 – A B&M dive coaster with 8-across seating

Friction

Friction is the resistance that one surface or object encounters when moving over another. Unlike drag, dry friction is nearly independent of velocity and acts tangent to the surfaces. Friction in the wheel bearings and between the car wheels and the track results in heat and transfers kinetic energy away from the car. Other sources of friction:

Wheel Bearing Clearance – Correct assembly of the wheels reduces internal friction and heat. The wheels need to be perfectly aligned. If you've ever made a pinewood derby car, one of the first things you should do to make your car faster is by reducing the amount of energy lost due to friction: make sure the wheels are straight and that the car rolls in a straight line. Same for roller coasters.

Wheel Bearing Lubrication – The amount and type of grease affects the speed of the train.

Track Lubrication – If the tack is lubricated the train will slow down less due to less frictional head losses. Track lubrication also prevents excessive wear to the rails and the wheels, especially those of wooden roller coasters. Maintenance workers at Merlin's Mayhem, a family suspended coaster by S&S at Dutch Wonderland, informed me they apply grease to the first third of the lift hill every morning. As the train passes, the wheels pick some up and distribute it around the track.

The application of lubricating grease is a balancing act that plays a vital role in controlling the speed of the roller coaster: too little grease speeds up the train, while too much slows it down, even causing it to grind to a halt mid-ride. Coaster operators may use a lithium-based grease which has the consistency of peanut butter at room temperature but changes to a liquid state as the wheels heat up. Adding grease means there is more grease to absorb the generated heat from the bearings, so the grease stays thicker in consistency. Thicker grease makes for a slower ride. But, by contrast, if there is less grease to absorb the heat, the grease itself heats up and thins out.

Rolling resistance – Rolling resistance, sometimes called rolling friction or rolling drag, is the force resisting the motion when a body rolls on a surface. Different wheel materials have different rolling resistances. Operators can choose different wheel material as well as wheel diameter size. More on this when we discuss wheels later.

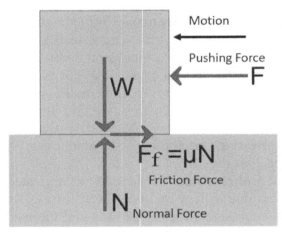

Figure 15 – Free-body diagram of friction forces

How does temperature affect roller coasters?

The lower the temperature, the slower the train goes. Why? The answer is in viscosity, the opposition to flow, which strongly depends on temperature. When you squeeze a bottle of syrup out onto your morning pancakes at room temperature, the syrup slowly drips out of the bottle. What happens if you heat up the syrup in the microwave first? It pours out much more quickly. That's viscosity. When a liquid heats up, its molecules become excited and begin to move. The energy of this movement is enough to overcome the forces that bind the molecules together, allowing the liquid to become more fluid and decreasing its viscosity.

So, all the lubrication in the wheel bearings, grease on the track, and so on, the resistance to flow decreases with increased temperature. Conversely, when the temperature drops, the lubricants are less effective at doing their jobs, thus affecting the speed of the train. Cold operating temperatures present challenges especially due to increased friction from grease viscosity. This can be countered by having heaters under the maintenance bay or station area to warm up wheels and grease on trains.

Time of day is another factor of influence, especially on a wooden coaster. This is usually directly related to temperature. In the morning, the trains have been sitting cold all night and haven't warmed up. Often, at times towards the end of the year, amusement parks will "deadweight" the coasters just to get them around the course. This is due to the temperature of the wheels — if they're cold, the ride runs a great deal slower than if the ride has been running all day. It's because of this that enthusiasts like to ride at night; warm wheels and lubricants equal faster rides resulting in more intense feelings of airtime.

The B&M dive machine Valkyria is said to be designed to run in 21 deg F (-6 C). It has a top speed of 65 mph and heavy trains less impacted by resistance forces than lighter trains would be.

How does mass affect roller coasters?

Without considering air resistance, velocity is independent of mass. In a vacuum, a five-pound cannonball dropped at the same height and time as a twenty-pound cannonball will hit the ground at the same time. Why?

Potential Energy = Kinetic Energy
PE = KE
$mgh = \frac{1}{2} mv^2$
Then mass divides out so you're left with:
$gh = \frac{1}{2} v^2$

However, in the real world we do have to deal with resistance forces like air drag. The portion of kinetic energy lost to drag is less with a larger mass object. The larger the mass, the larger the momentum, and the more force you need to change it. When a feather and a cannonball are dropped at the same time in the real world, the cannonball hits first. Air resistance has a greater impact on the lightweight feather. Mass does not make a roller coaster go faster in terms of the maximum possible top speed but it does make it harder to slow down.

This is why amusement parks test roller coasters with dummies filled with water. The water dummies increase the mass of the train making it harder for the resistance forces to slow it down so it's less likely to get stuck. This is especially useful when breaking in a new ride where the conditions may lend towards greater resistance forces: cold weather, more friction between brand new components, wind, etc.

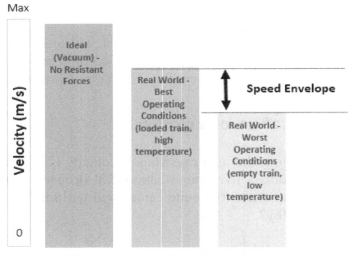

Figure 16 – Speed envelope example

Roller coasters operate within a speed envelop, or window. The low end of the window is the slowest speed the coaster can go yet still make it through the course. This is seen during unfavorable conditions such as an empty train running in low temperature. The high end of the window is the fastest speed the train operating in optimal conditions: fully loaded train, high temperature. Because the speed of the train varies but the curvature of the track remains the same, the gs experienced by riders will also vary. Trim brakes are added if the calculated gs exceed acceptable safety standards. Trim brakes can reduce the speed variance; making the speed envelop smaller so that every ride feels relatively the same regardless of the operating conditions (and has the added benefit of reducing wear and tear).

Some of the factors that influence speed can be controlled by the park, such as the number of guests allowed on the train, and when and what kind of lubrication to use. Other factors usually cannot be controlled, such as temperature, wind (unless built entirely indoors), and the natural lubrication effect of rain. These factors also play larger or smaller roles depending on the type of coaster: wood or steel. Track lubrication is primarily a concern of wooden coasters whereas wheel material is a bigger factor for steel rides. A roller coaster designer's goal should be to minimize the kinetic energy transferred to thermal energy as much as possible.

Free Fall Example

When Millennium Force opened at Cedar Point in 2000 it wasn't unusual to hear a coaster enthusiast claim the ride was running so fast that it was hitting 100 miles per hour. Is this even feasible? Let's check the math to find the final velocity at the bottom of the first drop given what we know.

Start with the known variables: Millennium Force's first drop is 300 feet or 91.44 meters (always make sure you're using the same units). The coaster is 310 feet tall but the distance the vehicle drops is only 300 feet. The cable lift operates at around 22 feet per second or 6.7 meters per second), so we'll assume that is the train's initial velocity at the top of the first hill. We also know g is 9.8 meters per second squared.

From the definition of velocity, we can find the velocity of a falling object and its altitude using these formulas:

$$v = v_0 + g \cdot t$$
$$h = v_0 \cdot t + 0.5 \cdot g \cdot t^2$$

We don't know t, the time it will take for the train to reach the bottom of the first drop. Since we know h, Vo, and g, we can solve the second equation for t then plug that value into the first equation to find the final velocity.

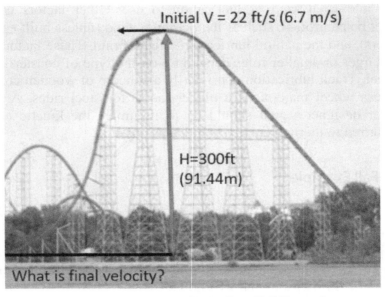

Figure 17 - Millennium Force Free Fall Example

Plugging in our known values:

$$91.44 = 6.7t + 0.5 \cdot 9.8t^2$$

Multiply both sides by 100:

$$91.44 \cdot 100 = 6.7t100 + 0.5 \cdot 9.8t^2 \cdot 100$$

Now simplify:
$$490t^2 + 670t - 9144 = 0$$

For a quadratic equation of the form $ax^2 + bx + c = 0$ the solutions are
$$x_{1,2} = \frac{-b \pm \sqrt{b^2 - 4ac}}{2a}$$

Inserting our values for a, b, and c:

$$t_{1,2} = \frac{-670 \pm \sqrt{670^2 - 4 \cdot 490(-9144)}}{2 \cdot 490}$$

The solutions to the quadratic equations are:
$$t = \frac{-335 + \sqrt{4592785}}{490}, t = -\frac{335 + \sqrt{4592785}}{490}$$

Select the positive value then insert t into the velocity equation:

$$v = 6.7 + 9.8 \cdot \frac{-335 + \sqrt{4592785}}{490}$$

The final velocity at the bottom of the 300-foot drop is 42.86 meters per second or 95.88 miles per hour. What does this mean? The free fall formula does not account for friction or any other energy losses. Therefore, in a vacuum the fastest Millennium Force could ever go is 95.88 miles per hour. No matter how hot it is or how much mass you add, it can never go faster than that without any additional means of propulsion. The giga coaster has been clocked at 93 miles per hour meaning Millennium Force is 97% efficient. Now if the conditions were right, with high temperature, wheel selection, grease selection, fully loaded train, the max speed might get a little closer to 95.88 but it cannot surpass that value without more height being added to increase the potential energy.

Chapter 4: Coaster Elements

Lifting Mechanisms

How do roller coasters reach their highest point? How do lift hills work? What angle are roller coaster lift hills? What types of lift hills are there? How has lift hill design changed over the years? What's the latest lift hill technology? In this chapter we'll cover all this and more.

What is a lift hill?

A lift hill is a mechanism used to transport a roller coaster car or train up a hill to an elevated point. As the train is pulled to the top, it is gaining potential energy, or stored energy. Once released, the acceleration due to gravity makes the vehicles coast back to the station. The lift hill is usually the first and tallest hill on a roller coaster (but not always).

Most roller coasters only have one lift hill, but they can have as many as required. Big Thunder Mountain at Disneyland Paris has four lift hills (The Ultimate at Lightwater Valley uses eight different chains, but five are used like friction wheels in flat segments – I would only count three as lift hills). Of course, more lift hills equal more expense with the structure required to hold them up, motors, chains, catwalks, and so on.

The higher the lift, the greater the amount of potential energy gained by the train. This is shown by the equation for potential energy: **PE=mgh** where PE is potential energy, m is mass (kg), g is acceleration due to gravity $9.8 m/s^2$, and h is the distance about the ground (m).

Because mass and gravity are constant for the train, if the height of the train above the ground is increased, the potential energy must also increase. Thus, gravitational potential energy is greatest at the highest point of a roller coaster. As the train accelerates down the hill, potential energy is converted into kinetic energy (and some energy is transferred to heat due to friction as discussed). A coaster with only a lift hill depends purely on the height of the lift for its energy. Launch coasters obtain their max energy through other means not directly related to the ride's height. Some roller coasters, like TMNT Shellraiser at Nickelodeon Universe, have both a lift and a launch.

In the early days of roller coasters, lift hills had a relatively shallow angle of ascent, angled between twenty and thirty degrees. There were very few variations, and all looked nearly identical. With today's technology, lift hills can have any angle of ascent. Maurer Sohne builds the craziest looking lift hills including a handful that even go upside down! On G-Force at Drayton Manor (now defunct) riders are pulled up the first half of a vertical loop. After being released from the weird lift the vehicles finish the loop and complete the rest of the coaster's twisted circuit.

Figure 18 – The bizarre upside-down lift of G-Force

Roller coasters must have a way to safely evacuate riders if the lift mechanism fails or the vehicles become stuck. These come in different varieties including evacuation stairs, carts, or the ability to reverse the lift to take the cars back to the ground.

Factors that influence the lift hill design and what type of lift is used include:

Space available	Method for rescuing riders
Noise	Materials used
Budget	Force on the components
Speed required	Rider Comfort
Height of the lift hill	And so on…

With improved engineering tools, lift hill support structure has evolved to become more efficient. Thicker track spines result in fewer support columns. The chain return is now usually located inside the spine of the track rather than a separate component. Orion is only 57 feet taller than Diamondback but in the eleven years between you can really see how the lift hill design has improved.

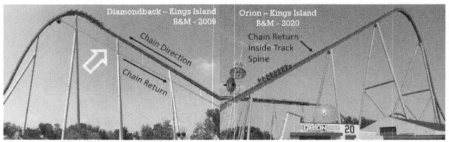

Figure 19 – Diamondback versus Orion lift hill design

The tallest lift hills on coasters today use "arch shape" support strategies to minimize the amount of structure needed. Arches are strong and press the weight outwards towards the main supports. Rocky Mountain Construction has become famous for their truss lift structures with inverting stall elements connecting the sides of the arch, as seem on Goliath, Zadra, and AerieForce One.

Figure 20 – See the arch shape on Intimidator 305, Fury 325, and Goliath?

The tallest lift hill in the world is Carowinds' Fury 325 at 325 feet tall (99.06 meters). From a rider experience perspective, lift hills build anticipation for what is to come and can offer incredible views.

	HEIGHT (ft)	TOP TEN TALLEST COASTERS	TYPE
1	456	Kingda Ka	Launch - Hydraulic
2	420	Top Thrill Dragster	Launch - Hydraulic
3	415	Superman: Escape from Krypton	Launch - Magnetic
4	367	Red Force	Launch - Magnetic
5	325	Fury 325	Lift - Chain
6	318	Steel Dragon 2000	Lift - Chain
7	310	Millennium Force	Lift - Cable
8	306	Leviathan	Lift - Chain
9	305	Intimidator 305	Lift - Cable
10	287	Orion	Lift - Chain

Lift hills come in all different types, shapes, and sizes. Except for magnetic (LSM or LIM) lift/launches, every type of lift hill requires a physical component on the vehicles to contact a component on the lift hill. Now let's look at all the different styles of lift hills there are today.

The **chain lift** is the traditional method of pulling a coaster car to the top of the tallest hill and has been in use for more than a hundred years. An electric motor powering the chain pulls the vehicle to the top of the tallest hill where its potential energy is now great. A catch on the bottom of the train called a "chain dog" or "chain pawl" engages a chain running in a trough fixed to the center of the track.

Why "chain dog"? A pawl is "a pivoted catch designed to fall into a notch on a ratchet wheel so as to allow movement in only one direction". A "dog" is defined as "a click or pallet adapted to engage the teeth of a ratchet-wheel, to restrain the back action; a click or pawl". Or "any of various mechanical devices for holding, gripping, or fastening something, particularly with a tooth-like projection." The terms are often used interchangeably.

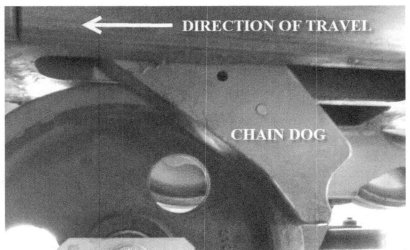

Figure 21 – Chain dog on a B&M hyper coaster

Sometimes, in steel coasters, the chain is guided in a plastic profile with two shoulders to prevent it from rubbing against the steel channel and getting worn out. The chain has pins on either side of the links and runs inside a channel like this: [__]. If the highly tensioned chain were to snap, the pins prevent the chain from flying back and hitting you in the face. The factor of safety for the chain is defined as the ultimate tensile strength of the chain divided by the maximum steady state tension. Chains in the primary load path that pass around sprockets of wheels shall have a minimum factor of safety of six.

Nearly every coaster is equipped with automatic lubrication units to periodically apply grease to the chain to reduce friction enabling it to easily slide in its trough. The advantage of using a standard lift chain on a roller coaster versus a launch or other system is to ensure reliability and stability. Instructions for how to care for the chain should be provided in the maintenance instructions.

Figure 22 - Chain lift hill components

RMC I-Box Track ↑   Chain Sleeve ↗   ↑ Chain   ↑ Anti-roll back teeth   Evacuation Stairs ↑

Modern roller coasters may use variable speed DC motors versus older chain lifts powered by AC motors. With AC motors, the lift motor is turned on in the morning and the motor runs at the same speed all day. It's either on or it's off. Try listening for this the next time you're standing in line for an old, wooden coaster.

Retrofitted old rides and new coasters come with DC motors utilizing a variable frequency drive (VFD). A computer controls the speed of the chain and reduces it to a crawl when no trains are present and then speeds up as the vehicle approaches. The speed can be dialed in exactly, so you have a smooth engagement of the vehicles to the chain. The purpose of varying the chain speed is to reduce energy consumption, reduce unnecessary noise, save on wear, and tear on the chain, and provide a smooth engagement between the chain and train. Sometimes the lift hill is also one of the block zones on a coaster where the vehicles can be safely stopped and evacuated.

Variable speed lift chains used to provide smooth and comfortable engagement with the train are also used for storytelling - an example of ride and show systems working together. The perfect example of this is Mystic Timbers at Kings Island when the lift slows down so the riders can hear a "Don't go in the shed!" warning.

On most roller coasters, vehicles are only permitted to travel in one direction up the lift hill. The clack-clack-clack sound heard as a roller coaster train ascends the lift hill is due to a safety feature known as the anti-rollback (ARB) dog. This device locks into a step arrangement mounted on the lift hill and, in case of chain failure, will hold the entire train safely in place. Depending on the length of the train, most coasters are usually equipped with three anti-rollback safety ratchets so that a possible reversal of the train is limited to approximately a third of the pitch, or distance between, the safety rack's teeth.

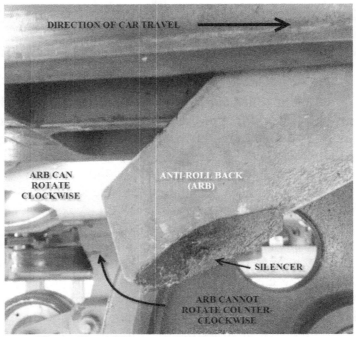

Figure 23 – ARB on a B&M hyper coaster

Many theme parks nowadays desire to keep their rides as quiet as possible, either to keep their neighbors happy or so as not to ruin the thematic experience. As a result, coaster manufacturers use "silencers" such as coating the end of the ARB with urethane or plastic caps as a simple and inexpensive way to dull and deaden the metallic clanking sound. Others use more complicated mechanisms with rails and axles or even magnets.

Roller coaster engineers need to figure out what strength of chain or cable they need, how powerful the motor must be to pull up a loaded train plus the weight of the chain or cable, how often the chain needs to be replaced, and so on. A ride engineer may study dozens of product catalogs and have countless talks with potential suppliers before the right chain is found for their thrill ride. Factors such as strength and cost must be considered. Breaks do occur after years of service and in taller rides there are catches in the chain trough to keep the chain from sliding all the way down to the ground into a giant, messy pile.

There's at least one coaster where the lift hill is purposely designed to let the vehicles roll back down them for extra thrill - Big Grizzly Mountain Runaway Mine Cars at Hong Kong Disneyland uses a catch car chain lift system that releases the vehicles to roll backwards, simulating a broken lift as part of the ride's storyline.

A relatively new variation on the chain lift is the **cable or elevator lift**. This system allows for a faster and steeper lift hill, which is often quiet because the anti-rollback devices are electromagnetically disengaged by the passing train and automatically close after it passes. The cable is connected to a catch car that rides on its own guide in the middle of the track. The catch car attaches underneath the vehicle so as the cable is wound up on a giant drum the train is pulled to the top of the lift hill.

Millennium Force at Cedar Point was the first modern roller coaster to use a cable lift system. It can pull the nine car trains up its mammoth 310-foot-tall lift hill in just twenty-two seconds. The cable lift innovation was a key reason why Intamin was chosen as the manufacturer over a competitor who had proposed a space-consuming chain lift hill in their failed bid to build the coaster.

Catch car lifts are primarily used on boomerang style roller coasters as once they pull the coaster train to the top, the catch car releases allowing the train to plummet backwards down the same track it was just pulled up.

Also called friction or booster wheel lifts, **tire drive lifts** are quieter than chain lifts but may also have a harder time operating in wet weather as the system relies on two tires pushing against a fin attached to the underside of the vehicles. To protect against slippage in wet weather is all about the coefficient of friction. Different materials and different textures on the material can be changed to improve the interface between tires and vehicle. The number of drive units can also be studied: is it better to have a smaller force on more drive units or a large force on fewer units? The aforementioned Zambezi Zinger uses a tire drive system in a spiral lift configuration.

Some roller coasters use **vertical lifts** to gain their greatest potential energy. These special elevators typically lift a piece of the track straight up where it connects to the big drop or other following elements. This arrangement is mainly used to save space or to provide a unique ride experience on small to mid-sized rides. The Lost Coaster at Indiana Beach in Monticello, Indiana is a perfect example of a ride that utilizes a single platform vertical lift.

To improve efficiency and reduce the time waiting for the platform to come back down, there is another variation called the **double platform vertical lift** where there are two platforms connected on a loop, one starting at the top of the tower and the other at the bottom. They pass each other in the middle. This arrangement is seen on taller rides to help improve capacity, including Intamin water coasters like Divertical at Mirabilandia.

Electric spiral lifts are special in that the vehicles themselves contain small electric traction motors to pull the trains up the hill. The cars engage with an electrified rail on the center of the lift track that provides power to the motors. To keep the weight down the motors, have to be small so it is impractical to climb steep lifts, thus the lifts of this type are in an upward spiraling helix to maximize real estate and minimize the grade of the ascent. This system employs anti-rollback devices just as any other roller coaster and once at the top of the lift the train disengages the electrified rail and gravity takes over. Whizzer at Six Flags Great America uses an electric spiral lift.

There's another lift hill system called the **Push Spiral Lift** in which the track is also configured in an upward spiral. However, unlike the electric lift there are no motors attached to the cars. In this case the vehicles are literally "pushed" up the spiral lift hill. A central rotating structure inside of the spiraling track contacts a wheel on an arm attached to the vehicle. As the structure rotates the wheeled arm rides up the structure until it reaches the top where the track pulls away from the rotating structure and gravity takes over again. This technology is found on Zamperla's Volare flying coasters, three of which exist in the North America, Soarin' Eagle at Scream Zone (formerly the Flying Coaster at Elitch Gardens), Superflight at Rye Playland, and Time Warp at Canada's Wonderland, as well as Puss in Boots' Giant Journey at Universal Studios Singapore.

Another highly unusual device to get the coaster vehicle to its highest point is the **Ferris wheel lift**. One of these was built at Freestyle Music Park (now closed and formerly called Hard Rock Park) in Myrtle Beach, South Carolina. A single car was rolled out of the station and onto a short piece of track attached to a rotatable ring fixed to the inside of a Ferris wheel. The

Figure 24 – A Zamperla Volare Spiral Lift

Ferris wheel would complete half a rotation while the track segment stayed upright the entire time thanks to the rotatable ring. After reaching the apex the car would be pushed off at the top into a high-speed coaster circuit before returning to the station. Round About / Maximum RPM no longer exists, and this was the only Ferris wheel lift ever built due to the low capacity and technical issues associated with always getting the track segments to perfectly align.

A roller coaster with a sloped track like a traditional lift hill but uses launch technology (like magnets or friction wheels) to rapidly ascend to the top with some additional kinetic energy is referred to as a **lift/launch**. Examples are Maverick at Cedar Point, Lightning Rod at Dollywood, and Incredible Hulk at Universal's Islands of Adventure.

A People-Powered Coaster

If you travel to Wales in the United Kingdom, you'll find one of the most energy efficient coasters in the world. The Green Dragon at GreenWood Forest Park is the world's first people-powered roller coaster. It's based on an old-fashioned inclined railway system developed for mines and quarries intended to move heavy coal down steep slopes without having any external energy source. Here's how it works:

Riders climb up to the top of a hill level with the coaster's empty station where they board a funicular (or tram). The tram descends the hill on a track under the weight of the passengers. Through a system of cables and pulleys, the empty coaster train (which weighs less than the combined weight of the tram and passengers) is lifted to the top of the hill on a detached section of track. The passengers then disembark the tram at the bottom of the hill and must climb another walkway back up to the loading station to board the coaster train. The empty track on which the train was sitting uses gravity to descend back down the hill to await the coaster train at the exit platform, pulling the tram back up to the station level in the process.

It's all about energy transfer: When the passengers climb to the top of the hill, they gain potential energy. As the tram descends the potential energy is converted into kinetic energy to pull the train back up the hill.

Lift Energy Example

How much energy does it take to pull a coaster train to the top of the tallest hill? A motor operating a chain does "work" to lift the coaster vehicle to the top of the lift hill. In this example, assume a 200 horsepower (HP) motor. One HP is approximately 746 watts, so 200 HP x 746 watts = 149,200 watts. Divide by 1,000 to get 149.2 kilowatts. To compute the energy, we need to know how long the power is needed. Assume it takes one minute of time from the moment the train hits the lift to release at the top of the hill. One minute is one sixtieth of an hour. Thus, 149.2 / 60 = 2.48 kilowatt-hours of energy are required to lift the train to the highest point.

Next, calculate the cost to operate the lift hill motor for a month in the summer. The theme park is open twelve hours a day for thirty days in the month, or a total of 360 hours. Multiply kilowatts by hours by dollars per kilowatt hours to get the total cost per month. Assume the cost of energy is $0.15 per kilowatt hour.

149.2 kw x 360 hours x 0.15 $/kwh = $8,056.8 for the month. That's over $8,000 to operate one motor on one ride for a month!

Launch Systems

Launch systems are a thrilling alternative to the traditionally slower lifting mechanisms. Apart from the time required, other limitations like height restrictions or confined real estate may prompt a park to choose a launch system versus a gigantic, space eating lift hill. Potential energy is stored as electricity or compressed air before it is transferred very rapidly to the train via cables, tires, or magnets and converted into kinetic energy. Types of launch systems include electromagnets, pneumatics, hydraulics, flywheel, catapult, and friction wheel.

Launched coasters have always required a very high amount of power for the launch, which either required an expensive high-capacity electrical service which is drawn heavily on when you launch, or the use of a device to store energy from a lower capacity service. The job of the roller coaster engineers is to get energy from a power grid or storage source and transfer it to the coaster in the most efficient way possible. Potential energy is commonly stored as electricity, hydraulic fluid, or compressed air before it is transferred very rapidly to the train via a propulsion system and converted into kinetic energy. How much energy are we talking about here? For example, the average lift hill motor could use approximately 200 amps for sixty seconds to lift a train to the point of release at the top of a hill. Compare that to the high energy launch systems that may require 4,000 or more amps for approximately five seconds of launch time!

First, let's talk about the difference between permanent magnets and electromagnets. A permanent magnet means exactly what the name says, it always has a magnetic field and will always display a magnetic behavior. An electromagnet is made from a coil of wire which acts as a magnet when an electric current passes through it and the polarity can be flipped.

Electromagnetic propulsion uses strong electrical impulses to attract or repulse magnetic fins attached to the vehicles. The stator and rotor are laid out in a line (as opposed to a torque or rotation) which produces a linear force and contains no moving parts. This propulsion system is popular because it offers very precise control of speed and in some cases the direction. The energy output is a sine wave where we can control how far apart the peaks are. For constant speed, the peaks will be a constant distance. To accelerate, the peaks will get closer together.

There are two types of electromagnetic propulsion used on roller coasters: linear induction motors (LIM) and linear synchronous motors (LSM).

Linear Induction Motors use multiple sets of high-powered electromagnets secured to the track. A gap is left in-between each set. Alternating current (AC) is applied to the magnets to create a magnetic field. A metal fin, usually made of aluminum, attached to the bottom of the train passes through the gap in the magnets while the magnetic field creates a wave for the fin to ride and propels or slows the train. The real advantage of LIMs is that they don't require the launch system to know the exact position and speed of the train. The train is basically riding on a magnetic wave which is traveling at a fixed speed down the launch track, and the train will accelerate to catch up to the speed of the wave the same way that people get pushed to the shore in a wave pool. In 1996, the Flight of Fear at Kings Island became the first roller coaster to use LIMs.

Linear Synchronous Motors use the basic magnetism theories of attraction and repulsion. Strong, permanent, rare-earth (those which come out of the ground magnetized) magnets are attached to the train. As with LIMs, secured to the track are electro-magnets. When the train approaches one of the track-magnets, the track-magnet is set to attract the magnets on the train, pulling the train forward. After the train passes over the track-magnet, the track-magnet is reversed to repel the train magnet, pushing the train down the track. Multiple sets of electro-magnets on the track must be fired in sequence, switching polarity very quickly using computers and electricity, to propel the train to top speed. LSM systems require many sensors to know exactly when the train passes each magnet so the polarity can be reversed at the right time.

How are LSMs efficient? Taking a step back for a minute, a magnet is any object that produces its own magnetic field. Every magnet has two poles, a north pole (N) and a south pole (S). Opposite poles attract and like poles repel. So, to not waste any energy in our launch system, when the poles line up (S-S, N-S, N-N) we want the voltage to be zero. The max voltage should occur halfway between. If there's no power when the poles are aligned, you're relying on momentum to get you to the next power input zone. To compensate, you have more than one permanent magnet attached to the underside of the coaster car so while one is deenergized the other can be energized. It's the same principle how the pistons on either side of a steam locomotive are offset.

To calculate magnetic force:

$$F = (n \times i)^2 \times \mu_o \times \left[\frac{a}{(2 \times g^2)}\right]$$

F = Force
i = Current
g = Length of the gap between the solenoid and a piece of metal
a = Area of electromagnet
n = Number of coils
$\mu 0$ = electromagnetic constant = 4 x PI x 10-7

An example of a LSM powered ride is Superman: The Escape at Six Flags Magic Mountain in Valencia, California. It was also the first roller coaster to reach speeds of one hundred miles per hour (though the track is not a complete circuit).

LIM and LSM drives are elegant, robust, and do not require extensive maintenance due to the lack of moving parts. However, they do draw a large peak power. There's only so much current that can be pumped into a stator before it reaches it limit. Temperature goes hand-in-hand with this as well because the harder a single stator works the hotter it's going to run. Both forms of linear motors operate with an air gap between the stator and rotor. The linear synchronous motor can operate with a larger air gap due to the fixed magnetic field in the reaction plate. The air gap between the stator and the rotor is directly proportional to the efficiency of the LIM or LSM.

It used to be that the maximum speed of the launch is limited by several factors. For instance, land available for the launch track limits the number of stators and ultimately how fast the top speed can be. However, a popular way to save space is by using a "multi-pass" launch where the same set of LSMs are used three or more times in succession launching a train forwards, backwards, and forwards again, gaining speed with each pass through of the electromagnets. Because of their versatility and efficiency, LIMs have essentially been replaced by LSMs. The last major new "ground-up" roller coaster to use LIMs was Big Grizzly Mountain Runaway Mine Cars at Hong Kong Disneyland in 2012.

Lift hill motors can be left on, spinning without a load on them other than the chain without a problem. The same cannot be done with linear induction motors. If the motor is left on it could overheat and burn up. LIMs must be turned on just before the vehicle approaches and turned off as soon as it passes. This is accomplished via precision control systems, a critical component of any launch coaster.

Hydraulic launch systems utilize a hydraulically powered motor to wind up a cable very fast. First, there is a catch car, called a sled, connected to a cable which latches on to a mechanism attached to the underside of the coaster train called a "launch dog". The catch-car moves in its own track or "groove" in the center of the launch track. The launch dog is held up inside the train by two permanent magnets. When the train is ready to launch, an electrical current is passed through two coils of wire wrapped around the magnets. This produces a magnetic field pushing the launch dog down away from the car. The train rolls backwards down the slightly sloped launch track allowing the launch dog to slide into a groove in the catch car. At the end of the launch, the launch dog is pushed back up inside the car. This way, if the train were not to make it over the top of the hill and rollback there would not be a catastrophic collision between the launch dog and the catch car.

The hydraulic motor is located at one end of the launch track and the waiting train at the other. Think of it like a giant fishing pole that reels a train in super-fast before being released. Here's how it works:

Hydraulic fluid is pumped into several different hydraulic accumulators (energy storing devices), comprising of two compartments separated by a piston. As the incompressible hydraulic fluid is pumped into one compartment, a gas in the other compartment is compressed. The nitrogen in the accumulator tanks starts to go under pressure as hydraulic oil is pumped into the tanks.

Once the nitrogen is compressed to an extremely high pressure, the pumping stops and the nitrogen goes into a cylinder block.

At launch, the fluid under pressure from the accumulators is used to drive either sixteen or thirty-two hydraulic motors connected to an internal ring gear. The power from all the motors is transferred to the giant cable drum by a gearbox. The cable drum spins, rapidly winding the cable attached to the sled hooked under the train, accelerating it in a matter of seconds. The train is released from the sled and speeds through the rest of the layout, but the sled and cable drum must decelerate rapidly by eddy current brakes and then return to their initial starting positions via the cable system to launch the next train.

Typically, there are two sensors mounted to the top of the tallest hill immediately after the launch track. The distance between the two sensors is known so the control system can take that value and divide it by the time it takes the train to get from one sensor to the other. This gives the computer the train's speed going over the top of the hill. For every single launch, the information is recorded and plotted as a performance curve. The computer takes the average speed of the three previous trains and compares it to past launches to determine the power to give to the hydraulic motor. This way, if the first three

Figure 25 – A Hydraulic Motor

trains are filled with swimsuit models and the fourth train is carrying football players the power is enough to get the car over the hill.

Hydraulic launch systems are considered capable of giving a far greater and smoother acceleration than current electromagnetic propulsion styles. The acceleration from a hydraulic launch remains nearly constant throughout the entirety of the launch. However, the number of moving parts makes this system generally less reliable than magnetic systems that contain no moving parts. Hydraulic launch systems have the highest power and are compact but the whole cable drive part is not trivial due to the forces and speeds.

The first hydraulic launch coaster was Xcelerator at Knott's Berry Farm in 2002 with a zero to eighty-two miles-per-hour in 2.3 seconds. The last new hydraulic coaster opened in 2010 and to this day remains the world's fastest: Formula Rossa at Ferrari World in Abu Dhabi, United Arab Emirates achieves an amazing speed of 150 mph (240 km/h) in 4.9 seconds! Hydraulic launches have largely been replaced by more versatile LSM launch systems. The world's tallest roller coaster, Kingda Ka, hydraulic motor can produce a peak power of up to 20,800 hp (15.5 MW) for each launch. One thing to keep in mind is that although hydraulic launches have a lower peak load requirement since the electricity is drawn at a constant rate all day long to continuously pressurize the accumulators, they still use more energy overall because it is a less efficient system due to all the moving parts and mechanical losses.

Figure 26 – Kingda Ka, world's tallest roller coaster

Pneumatic launches are very similar to hydraulic launches; the major difference being air is compressed instead of nitrogen gas or oil. Acceleration is nearly constant, but the resulting loud noises caused by the whooshing air can be an annoying issue for theme parks. Pneumatic systems are technically simpler than hydraulics but are also less powerful. Dodonpa at Fuji-Q Highland in Japan can launch passengers from 0 to 106.9 mph (171 km/h) in 1.8 seconds using compressed air!

Electro-magnetic, pneumatic, and hydraulic launch systems are the popular types of acceleration technologies used today as they are much more efficient and powerful over the older and less common launch systems. **Flywheel launchers** utilize a large flywheel (a device used to store rotational energy) that is spun at high speeds and is attached to a cable that propels the train forward. **Friction wheel launches** have a launch track that consists of a series of horizontal tires that spin in opposite directions. They pinch the metal fins on the underside of the train, much like a baseball pitching machine.

The Incredible Hulk coaster at Universal's Islands of Adventure in Orlando, Florida is a perfect example of a friction wheel launch coaster. However, a great deal of power is needed to propel the 32-passenger train from 0 to 40 mph (64 km/h) in 2.0 seconds. The launching sequence for the Hulk hammers the local electrical grid with 6,000 amps, so much energy, that if it depended on the power supply that the rest of Orlando uses, lights all around the city would dim each time it catapulted another train of screaming guests into space. To achieve the brief but very high current required to accelerate a full coaster train to max speed at an uphill angle, Universal engineers built a separate power plant just for the Hulk that sucks electricity from the city at a steady rate and stores it in four massive, 10,000-pound flywheels. Without these stored energy units, the park would have to invest in a new substation or risk browning-out the local energy grid every time the ride launches.

Catapult launches involve a large diesel engine or a dropped weight to wind a cable to pull the train until it accelerates to its full speed. In 1977, Kings Dominion in Doswell, Virigina opened the first launch coaster by means of a weight drop. A massive, 90,000-pound steel-reinforced concrete weight was attached to one end of a 50mm diameter steel cable that wrapped around a giant winch, known as a bull wheel. The wheel was located under the station and the other end of the cable was connected to a pushcart. When the operator hit the start button, the brake on the wheel released and the weight dropped inside of its silo. This action caused the pushcart to accelerate and push the train to a speed of 50 miles per hour. Walibi Belgium's Psyké Underground is a Schwarzkopf shuttle loop coaster with a flywheel launch. Originally opened in 1982, for the 2013 season the flywheel launch was replaced by LIMs.

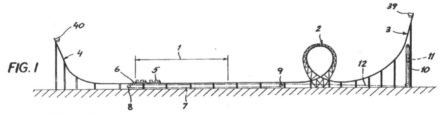

Figure 27 - Anton Schwarzkopf's patent for the Shuttle Loop

Vertical Loops

There is something quite remarkable about being able to defy gravity and part of the charm of roller coasters is they let humans do exactly that. The vertical loop is one of the most common, yet thrilling elements found in steel roller coasters. Very few sensations can match the excitement of dangling upside down at a great height without the danger of falling out.

Early roller coaster loops, including the first one, a 13-footer built in 1846 in Paris, were simple circles. To make it all the way around without stalling, coaster cars hit the circle hard and fast, shoving rider's heads into their chests as they changed direction with a sudden snap that occasionally broke bones.

Coasters 101: An Engineer's Guide to Roller Coaster Design

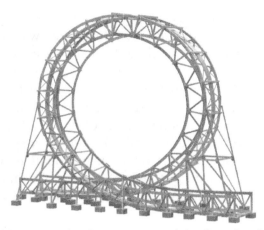

Figure 28 – Rendering based on one of the first circular loops

The shape of the loop determines how much force is felt at any location along it. The intensity of the acceleration force is determined by the speed of the train and the radius of the track, as seen in the formula: $F = ma = m\frac{v^2}{r}$. At the bottom of the loop the acceleration force pushes the riders down in the same direction as gravity. At the top of the loop, gravity pulls the passengers down towards the ground, but the stronger acceleration force pushes upward towards the sky, keeping riders in their seats.

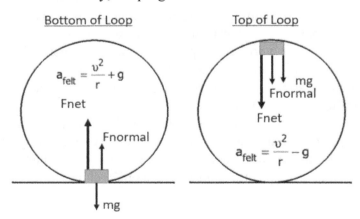

Figure 29 – Vertical loop free-body diagram

Roller coasters are forced through the loop by the track applying **centripetal force** to the cars. Centripetal force keeps an object moving in its circular path and is the same force that prevents water from falling out of a bucket that is swung upside-down on a string. In Figure 27, the net force is the centripetal force where $F_{net} = m(V^2/r)$. The normal force, exerted by the track, is the force felt by the passengers. At the top of the loop the force felt equation is changes because the centripetal force and the force of gravity point in the same direction.

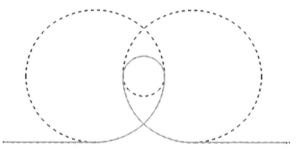

Figure 30 – Using different size circles for loops

Using circles of different radii, as shown in Figure 28, is better but still doesn't solve all the issues. The disadvantages of circular loops are not limited to the maximum g-force at the bottom: entering the circular loop from a horizontal track would imply an instant onset of the maximum g-force, which could be an instantaneous g increase from 1 to 5 gs. The rate of g force onset or change is also important to consider to keep passengers safe.

Vertical loops were finally made safe and comfortable in 1975 by Werner Stengel with the teardrop shaped loop. He ditched the perfect circle and designed a loop with a radius of curvature that decreases as the vehicles are turned upside down. This way the g-forces at the top of the loop can be kept much closer to those of the bottom, resulting in a smooth and enjoyable experience.

Coasters 101: An Engineer's Guide to Roller Coaster Design

Figure 31 – Rendering of an early teardrop shape loop

The decreasing radius of the track is designed using an Euler spiral or clothoid (or klothoide and pronounced 'clockoid') configuration. Clothoids are frequently used in railways, road building, and highway exits. A driver keeping constant speed in a clothoid segment of a road can turn the steering wheel with constant angular velocity; the Cornu spiral has the property of the radius of curvature being inversely proportional to the distance from the center of the spiral. While the upper part of the vertical loop could still be a half circle, the lower part has a completely different shape. It's part of a "Cornu spiral", where the radius of curvature increases as you get closer to the ground. Parts of clothoids are also used to connect parts of tracks with different curvatures. The clothoid shape leads to a slower onset of lower forces on the body, leading to a much safer ride for passengers (and no broken bones). The first roller coaster with a modern (clothioid) loop was Revolution at Six Flags Magic Mountain in 1976.

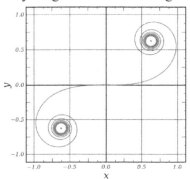

Figure 32 – Cornu spiral

There are other methods using mathematical formulas to create safe loops including keeping constant centripetal acceleration, clothoid, and keeping a constant g force. Many loops are a mix and match of different formulas. Some manufacturers are known for specific kinds of loop and other inverting element shapes. Coaster elements are not patented or protected, and most are used widely among all manufacturers. If you want to get more technical on the physics and mathematical formulas used by roller coaster engineers in creating the perfect vertical loop, watch the YouTube video *The Real Physics of Roller Coaster Loops* by Art of Engineering.

Ironically, some coaster manufacturers are now creating loops with more circular or elongated shapes to produce hangtime. Hang time is when a coaster vehicle is going through an inversion with just enough speed to not stall but not quite enough force to keep passengers in their seats, resulting in hanging against the restraints.

Today there are four roller coasters tied for the world's largest vertical loop at 160 feet (48.8 meters): Flash, Hyper Coaster, Do-Dodonpa, and Full Throttle. The world's highest inversion above the ground is the 197-foot dive drop on Kennywood's Steel Curtain. The most inversions in a track on a single coaster is The Smiler which will flip you upside down an absurd 14 times.

Figure 33 – Copperhead Strike's more circular loops produce hang time

Banking the Track Through a Curve

To help keep the forces on the riders under the acceptable limits, the roller coaster's track is banked whenever it negotiates a turn. If no amount of banking will keep the g-forces on the riders at a safe level, then the centerline of the ride must be adjusted accordingly.

There are a few different techniques to go about banking a curve on a roller coaster. In some earlier coaster designs, the banking on a curve was achieved by holding the inside rail level and raising the outside rail, which was rotated about the inside rail.

Figure 34 - Rotation about the track spine

You also had designs where the inside rail was lowered, and the outside rail was raised with the track rotated about the spine or backbone pipe.

Figure 35 - Rotation about a rail

In these types of arrangements, the center of gravity of the passenger is accelerated toward the center of the curve. This results in the passenger being thrown against the side of the car or against another passenger. As discussed earlier, the ride must be fun and not put too much strain on a rider. To correct for this acceleration, the track should be rotated around the center line or "heartline" of the passengers (which lies roughly just above the center of a human torso). Thus, the acceleration of the passengers to the inside of the curve gets greatly reduced. Remember, it's all about producing safe accelerations.

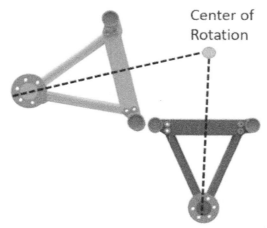

Figure 36 - Rotation about a centerline (not to scale)

There are generally two scenarios that you, as a designer, will want to avoid when designing a banked curve: overbanking and underbanking. At just the right bank for a coaster car's velocity, the car will not need any type of undercarriage or guide wheels to stay on the track (though they'll still be there for safety reasons). The rider will feel a normal force pushing him into the seat of the car. This is the optimal scenario to keep the car on the track, where no friction is needed.

Overbanking (or "overturning") occurs when there is too much banking for the car's velocity. In other words, the car could tip to the inside of the curve. The guide wheels hold the car on the track, with the rider feeling a force pushing him down into the inside of the curve. Friction is needed to keep the car on the track.

Underbanking is just the opposite; there is not enough bank for the car's velocity. The car can potentially tip to the outside of the curve and the guide and upstop wheels hold the car onto the track. The rider feels a force hurling him away from the center of the curve. Friction is also needed to keep the car on the track. In a few special situations the curves are intentionally underbanked by a small amount. This way, passengers feel a slight outward force around the curves, which they expect and naturally adjust to. This is done because an attempt to produce a perfect bank can accidentally result in an overbank due to a slower than expected train (which could possibly be slower from high winds, being half-empty, etc.).

There are a few coasters which deliberately bank their track to the outside of the curve to create a thrilling effect and unique feeling for occupants, with the restraints holding them in the vehicle. This effect works well in the dark owing to the element of surprise, with riders experiencing an eerily exhilarating and unnatural feeling as the coaster banks to the outside of the curve. This is called outerbanking (not to be confused with overbanking).

How is a perfectly banked curved produced? What angle should the turn be banked so that riders are squashed into their seats, but not tossed painfully to the side? The following equation is used:

$$\mathbf{Tan(\theta) = v^2/rg}$$

The Greek letter theta (θ) is the angle where no outside forces other than gravity are required to keep the car from sliding to the inside or outside of the curve. If the velocity, v, of the car is changing through the curve then the banking angle of the track must also change to compensate. R is the radius of the turn and g is gravity (9.8 meters per second per second). There are two components to the g forces felt: horizontal and vertical. To calculate the vertical g's felt for the ideal banked curve with no friction, use the following:

$$\text{g's felt} = 1/\cos(\theta)$$
$$-m*g + N*\cos(\theta) = 0$$
$$N = m*g/\cos(\theta)$$

In these equations, m = mass, N = normal force, and g = Gravity.

Curve Design Example

A new extreme, 90 mile per hour mega coaster opened at an amusement park to great fanfare. However, after a few weeks of operation the park received a large number of complaints from guests about the ride being too intense. Most of the riders found it to be more nauseous than fun and complained about the high g-forces experienced in the first turn following the giant first drop.

Additionally, the maintenance department had to replace wheels much more frequently on the super coaster than any other in the park. This metal monster is heavily loaded with extreme g-forces and heavy trains which result in significant wheel usage, especially in hot weather conditions. Wheels were lasting only several days during the summer with multiple modes of failure.

What could the park do to address these concerns? To fix the melting wheel problem an alternative wheel design with a different chemistry was used and the average wheel life increased by five times, resulting in significant cost savings. To reduce the force on the passengers as a temporary fix, a series of trim brakes were added to the first drop, significantly reducing the top speed of the ride. Fortunately, the ride's top speed could be reduced but still make it around the course without stopping. But adding trims to the first drop reduced the thrill level and the park could no longer market the ride at its top speed. The problem needed to be fully understood in order to come up with a long-term solution.

Well, the main issue wasn't the high speed in the first turn; instead, the issue was angular acceleration – the vertical force that the train, the wheels, and the riders experience while going around the curve. The equation in question is: $F = m*V^2/R$ ($[N] = [kg].[m].[s^{-2}]$). The main objective is to reduce the force (F) on the wheels and riders. There are two ways to reduce this force: either reduce the velocity (V) of the train or increase the radius (R) of the curve.

The temporary fix was to reduce the velocity but the long-term solution was to change the radius of the curve. The park determined that the g-forces in the turn were too high, so the track needed to be re-profiled and replaced (quite an expensive correction). Often, a ride's structure has already been manufactured by the time the data is sent to the company hired to verify all the design assumptions and force calculations are correct. The track was modified by increasing the curve radius which allowed the train to run faster through the curve (but at a reduced force). The result also means the coaster will be able to take the rest of the course faster because the train will come out of the curve at a higher speed, thus returning the ride to its original and intended glory!

Figure 37 – A sit down coaster with lap bars and on-board audio

Nick Weisenberger

Chapter 5: Vehicle Design

A key relationship in roller coaster design is the pairing or mating of the vehicle to the track. Track design is dependent on vehicle performance so before designing the layout a defined interface between the car and the track must be agreed upon. The customer and designers must clarify which elements and features they want the cars to be able to negotiate as well as what type of seating arrangement for the passengers, the minimum turn radius allowed by the vehicle design, and the minimum crossover height allowed by the track and vehicle.

A roller coaster train is a string of multiple vehicles connected together. Single vehicles, or cars, will have wheels at both the front and the rear of the vehicle. Multiple cars can then be hooked together to form a train using coupling rods and uniball joints that allow the vehicles to follow the twists of the track together.

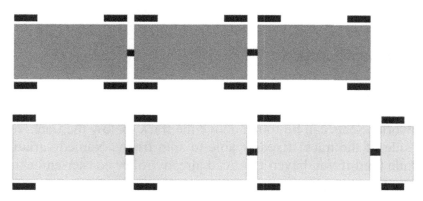

Figure 38 – A traditional train (top) versus a trailered train (bottom)

Most modern mega coasters use trailered cars to save weight. by reducing the amount of wheel assemblies needed. A trailered car has only a single axle in the front or rear and relies on the preceding car for balance. Another way to think of it is a three-point suspension with a single axel and a pivot in front. The first car, often called the lead car or zero car, does not have any seats and is only used for stability. They are usually heavily themed.

There are several advantages to having shorter trains including simpler design consideration and the ability to negotiate tighter elements. A shorter train allows tighter radii on the apex of the hills and drops. On a ride with a very long train, you can get a different ride experience depending on whether you sit in the front or the back. Shorter trains lead to more consistent forces. The consistency of the forces from front to back could be the difference in having a ride signed off for acceptable forces or not. When designing coasters that have wider cars, such as dive or wing coasters, the roll rate becomes more important since rolling creates additional accelerations that increase the further away from the roll center.

Why do some roller coaster trains have headrests and others don't? Safety standards dictate you must have a headrest above 1.5 g unless onset rate is less than 5 g/sec; then 2.0 g is permissible. For no head rest, the max duration of above 1.5 g is 4 seconds.

Seating Configuration

Just about any seating arrangement or configuration that can be imagined for a roller coaster has been attempted somewhere in the world. Seats can be found above the track, below the track, or to the side of the track; fixed or able to spin freely. Named variations include (and if you haven't noticed already roller coaster enthusiasts love to give everything a name):

- **4th Dimension**: Controlled rotatable seats cantilevered on each side of the track.

- **Bobsled**: Cars travel freely down a U-shaped trough (no rails) like a bobsled except on wheels.

- **Floorless:** The vehicle sits above the track but contains no floor between the rider's feet and the rails, allowing their legs to dangle freely.

- **Inverted**: The vehicle is fixed below the rails with rider's feet hanging freely and can invert upside down.

- **Laydown/Flying:** Riders are parallel to the rails, either on their back or stomach.

- **Motorbike:** Riders straddle the seats as if riding a motorcycle, jet ski, or horse. Also called straddle.

- **Sit down:** Traditional roller coaster with vehicles above the rails.

- **Spinning**: Seats can freely spin on a horizontal axis.

- **Standup:** Riders are restrained in a standing position, straddling a bicycle-like seat and shoulder harness that adjust to the rider's height.

- **Swinging suspended:** The vehicle hangs below the rails and can freely swing from side to side but does not invert.

- **Pipeline:** Riders are positioned between the rails instead of above or below them.

- **Wingrider**: The seats are fixed on both sides of the vehicle outside of the rails.

Wheel Design and Material Selection

Three types of wheels secure the vehicle to the track: **Road wheels** bear the load or weight of the train and are subjected to the biggest loads, which is why they're also called load wheels. The allowed vertical downward gs are greater than what is allowed for lateral or uplifting gs. Load wheels are usually also bigger in diameter as well.

Side friction wheels, or guide wheels, are mounted perpendicular to the road wheels on either the inside or the outside of the rail depending on the type of track. The side friction wheels on steel roller coasters are spring-loaded against the sides of the track. The wheels are forced to steer to follow the track. Conversely, wooden coaster wheels have historically been on fixed axles. There is a gap between the wheels and side of the tracks so the cars "shuffle" through the turns without exactly following the turning track. This results in more banging around and causing the track to wear because it eventually crushes the wood. New wood coaster trains have now introduced steering into their vehicles to help eliminate some of the roughness.

Figure 39 – A steel coaster wheel assembly

Side wheels can either be on the inside or the outside of the rails. Having side wheels inside the track allows lowering the heartline as trains can travel partly inside the track, and thus the track shape is easier to calculate by hand. This is partly why older steel coasters by Arrow Dynamics and Vekoma always had their side wheels inside the rails. The most recent major roller coaster in the US to use inside wheels was Kentucky Kingdom's Lightning Run by Chance Rides (who is also supposed to be opening another major inside wheel coaster in 2023).

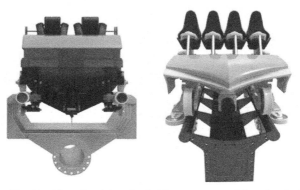

Figure 40 – Inside wheel track (left) versus outside wheels (right)

Most manufacturers have nowadays switched to only using tracks that utilize side wheels running outside the track. With advanced CAD technology it has become easier to set the heartline to any desired height and have the computer calculate the final shape of the rails. These benefits include easier maintenance access to the side wheels (resulting in a smoother ride experience), less used material in track manufacturing and a structurally sturdier track with less critical fatigue points. Companies like Vekoma have now switched to new outside wheel track designs (although they still use the old track for their mine train models so railroad-like cross ties can be added for theming).

Upstop wheels are placed under the rail to prevent the vehicle from coming off the track over airtime hills or if the vehicle were to stall in an inversion. Even if there are no accelerations that could cause lift-off, a safety device capable of withholding 50% of the fully loaded vehicle weight is required. Some coasters with no lift-up forces use small steel plates rather than upstop wheels.

Figure 41 – Wheels on a B&M hyper coaster

A typical wheel for a steel coaster is constructed by taking an aluminum hub and bonding a polyurethane tire to the hub's outside diameter. This entire "wheel assembly" is then connected to the axle through a bearing. The guide wheels are spring-loaded to ensure they always maintain contact with the rail. The unit with all six wheels attached is referred to as a wheel carrier.

One of the biggest limiting factors in building ever taller and faster roller coasters may be the wheels. To design the perfect wheel for a high-speed roller coaster, engineers must find the best combination of the following four main requirements:

1. Low rolling resistance
2. High load endurance
3. Smooth ride performance
4. High durability

Let's examine each of these requirements in more detail.

Rolling resistance is caused by the deformation of a tire at the point where the tire meets the surface on which it travels — in this case, the coaster's rails. The lower the pressure and/or the higher the force exerted on the tire, the larger the coefficient. For example, a non-deformable steam train wheel made of steel riding on a non-deformable steel rail has a very low rolling resistance, hence making it very efficient. To calculate the rolling resistance force:

$$\mathbf{FRoll = R \times M \times G}$$

R is the rolling resistance coefficient, m is the mass, and g is the acceleration due to gravity = *9.81 m/s²*. A typical rolling resistance coefficient value could be between 0.009 and 0.018 of the supported loads. On a roller coaster, energy losses due to friction must be minimized for the train to complete its circuit composed of complex maneuvers and dynamic inversions. The requirement of a low rolling resistance leads designers toward selecting a harder wheel material.

Several record-breaking roller coasters today plunge down mammoth four-hundred-foot drops, operate at speeds more than 120 miles per hour and can subject riders to forces more than six times that of gravity. The wheels not only carry the weight of the passengers and the vehicles, but they must also be able to do so up to six times their weight, possibly resulting in a load of 6,000 pounds of force on each wheel, with the train moving at a very high speed all the while. This again leads toward the selection of a harder wheel material.

The most important requirement from the perspective of a guest is making the ride as smooth as possible (more so for a steel coaster, although perhaps not so much for a woodie). Manufacturing large sections of track perfectly within tolerance without any imperfections is very difficult (and expensive) to achieve. Therefore, the wheels must absorb any deficiencies during manufacture or other conditions such as dirt or debris on the track. Rough rides are not only unpleasant for the park guests riding the coaster, but they can also cause damage to the vehicle over a period of time resulting in higher maintenance costs. The need to provide a smooth ride directly opposes the first two requirements by leading the designer to choose a softer wheel material.

A conventional roller coaster train may contain over one hundred wheels (12 wheels per car × 9 cars per train = 108 wheels per train) resulting in a significant portion of ongoing roller coaster maintenance being tied up in wheel replacement. Let's say we have a coaster with six-inch diameter wheels traveling at 70 mph. To calculate the revolutions per minute (or RPMs), first find the circumference of the wheel:

C = 2πr = 2×3.14596×3 in = 18.87 in

Convert to miles: 18.87 in×(1 ft/12 in)×(1 mile/5280 ft) = 0.000297 miles

RPMs = linear velocity/circumference =
70 mph/0.000297 miles=

234,968 revolutions per hour

234,968.02 / 60 min = 3,916 revolutions per minute

After running through the calculation, we see that the wheel will be spinning at nearly 4,000 rotations per minute, at least for a short duration during the circuit. It is an absolute requirement that the wheels last as long as possible. Amusement parks may be in the business of fun, but they are still a business after all and must be profitable.

Occasionally, the wheels will wear out or fail altogether. Sometimes an air bubble may develop in the layers of urethane. A blowout occurs when the wheel material's critical temperature is exceeded causing it to melt, maybe due to friction. In this case, the wheel must be replaced. Now, if a single wheel blows out on the coaster while it is in motion it can still make it around back to the station without incident where the wheel will then need to be replaced. Maintenance and storage tracks are now designed so that the load is taken off the main wheels when they're stopped for long periods of time. This also allows the engineers to easily inspect and replace the road wheels.

Another wheel failure mode is fatigue cracking, a normal wear pattern that usually occurs over a long period of time and is caused by high stress concentrations during use. Fatigue cracks typically do not pose a threat to the wheel until they begin to connect with each other, spread open, or reach a depth close to the hub. Since the rate of propagation of fatigue cracks differs between materials, applications and even wheel position, it is important to develop an inspection schedule to check for these possible failures.

Fatigued bond is a mode of failure that can lead to delamination and generally occurs over time as the bond is weakened through overloading or overheating. A fatigued bond can be a result of many different factors; however, no matter what the cause, it is easily discovered through proper inspection.

Steel roller coasters today generally use two main types of tire material: nylon and polyurethane. There are advantages and disadvantages for each type. The nylon wheel is a hard plastic while the polyurethane is a softer material. Nylon wheels vibrate a little more and put more wear into the track, making it a rougher ride but also results in a slightly faster ride. Polyurethane is a softer material and reduces the vibration, providing a smoother ride. However, it provides more friction and slows the ride down due to a higher rolling resistance.

Figure 42 – Two different steel coaster wheels

Urethane tires have certain disadvantages for use on wood coasters. They are less durable than the plain steel tires and are subject to damage from rolling over the track joints. Perhaps more important is the fact that owing to its flexibility the urethane tire can absorb a considerable amount of energy resulting in a slower ride. Yet several operators choose to utilize non-steel wheels on their wooden coaster, possibly due to noise limitations. These rides tend to be taller but shorter in overall length due to the higher rolling resistance of the softer tires.

The speed of a coaster can be affected by mixing and matching wheels, which parks may do to keep the ride running within its specified performance window. A new ride may use one type of wheels until it is tested and broken in, then will switch to the other kind of wheels. Many hours of testing may need to be performed to achieve a perfect balance between all the demanding requirements.

Another thing to consider is static discharge, especially for nylon wheels. Static electricity is an invisible field of protons, neutrons, and electrons found in anything that has potential to move. When an electrical charge such as static electricity builds up, it can have a violent discharge when the protons find the easiest path to the earth's surface, referred to as ground. Rolling friction will cause an imbalance of neutrons if the electrical charge cannot get to ground. To countermeasure this concern, engineers add straps to the vehicles behind the wheels. These mini mud flaps contact the rails to ground the train to protect the riders and any onboard electronics from damage by static discharge.

Restraint Design

The crazy inversions and extreme airtime of mega-coasters today has created a new level of fun for fanatics while breeding a new set of challenges for designers. This equates to designing new restraint systems to accommodate a wide variety of body types and sizes while maintaining comfort and safety. Restraint design is directly affected by the g-forces felt during the ride. ASTM International standards recommend the type of restraint required depending on the sustained accelerations, the riding position, or any other foreseeable situation like being stuck upside down. Every restraint falls into one of five categories:

- ❖ Class 1: unrestrained or no restraint at all.
- ❖ Class 2: a latching restraint for each individual rider or a latching collective restraint for more than one patron. Not required if there is sufficient support and the means to react to forces, like handrails and footrests.
- ❖ Class 3: restraint is required. A latching restraint for each individual rider or a latching collective restraint for more than one patron.
- ❖ Class 4: a locking restraint device for each individual patron.
- ❖ Class 5: A Class 5 restraint is required. Must be automatically locked. A class 5 restraint configuration can be achieved by the use of two independent restraints or one fail safe restraint.

For each class, the following need to be defined:

- ❖ Number of patrons per restraint device
- ❖ Final latching position relative to the patron
- ❖ Type of latching/locking
- ❖ Type of unlatching/unlocking
- ❖ Type of external correct or incorrect indication
- ❖ Means of activation
- ❖ Redundancy of latching device

As you'll see, terminology in design standards is very important. There is a difference between latching and locking. Latching is something a passenger can do themselves. Locking/unlocking is only accessible by the ride operator, not the patron. Restraints are locked either mechanically, hydraulically, pneumatically, or magnetically. Perhaps the most common use of hydraulics on modern roller coasters is in the restraint locking mechanisms. In case of failure, each patron is always secured with redundant safety features that work individually and can hold the rider safely on board.

From the rider's perspective, restraints come in all shapes and sizes but generally fall into two categories: lap bars or over-the-shoulder restraints (OTSR). Lap bars are U or T-shaped devices connected to the floor and rest on the rider's lap, securing them by the legs. Over-the-shoulder harnesses are mounted behind the rider and swing down over the shoulders. They also keep the upper part of the body straight and spine supported by the seat which helps the gs to be absorbed as designed. Both systems usually use a standard seat belt as a redundant fail safe. Recently, there is a trend of having "over-the-shoulder lap bars." These are lap bars that pivot about a point located behind the rider's head rather than from the floor.

Another feature the types of restraints have in common is only one degree of freedom, meaning movement is only allowed in one axis, a rotation. Having only one degree of freedom makes it easy for the control system to know the exact position of the restraint with respect to the rider. A go-no/go sensor tells the ride operator if the restraint is in the correct closed position before the coaster can be dispatched.

Clearance Envelope

While restraints are designed to keep passengers safe, there is still the opportunity for patrons to raise their arms in the air or stick their heads out of the side of the vehicle. A clearance envelope must be studied digitally during the design process to ensure there are no obstacles within reach. Customers losing their limbs would be bad for business after all. ASTM standards state: "the minimum patron model shall be based on Dreyfuss Human Scale 4/5/6, 7/8/9, SAE J833, or CDC 95th percentile, with an additional (extended) arm and length reach of 3 inches."

Once a roller coaster's structure and theming elements have been completed, a pull-through test will be performed to check the physical clearance envelope. A car with a special rig simulating the rider's farthest reach (plus three inches) will be slowly pulled through the entire layout to ensure all clearances are met.

Despite all the advanced tools at an engineer's disposal, mistakes still happen. Though

Figure 43 – A modified support

not confirmed by the park, Cedar Point's Millennium Force has a suspicious cutout in one of the first overbanked turn supports with a rectangular piece of bar stock welded to the back for reinforcement. At least one rider claimed to have hit their hand on the support before it was modified. How it was missed during clearance checks, we'll never know. The ride has operated safely for over 22 years. Every roller coaster has a minimum rider height requirement, but some now also have maximum rider height requirements as well.

Coasters 101: An Engineer's Guide to Roller Coaster Design

Vehicle Design Example

Roller coaster manufacturers have used almost every conceivable seating configuration, from inline seating to sitting up to ten seats across in a single row. It all goes back to design intent. What size and type of ride are is being built? What elements does it need to be able to maneuver? Consider the following example where there is a space limitation driven design change.

First, define "roll" as a rotation around the track centerline along the train's longitudinal axis running from the front to the back of the train. Pretend this coaster has a two-bench wooden coaster car with a four-foot wheelbase. It's been designed so the rear axle can swing about three degrees in either direction. That means the maximum allowable roll rate (without lifting any wheels) is three degrees per four running feet, or more usefully, about 16 inches of running length per degree. Meaning, to get up to a 90-degree bank from flat track, the train needs a minimum of 120 feet of track.

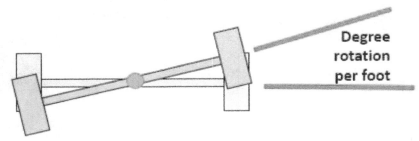

Figure 44 – Vehicle roll rate

Well, the roller coaster can't collide with the Ferris wheel so, the designers need to decrease the length of track it takes to get to a ninety-degree bank. Increase that roll rate from 3.0 degrees to 5.6 degrees per four running feet, or a mere 8.6" of running length per degree of roll by allowing one axle to rotate more relative to the other one. That reduces the length required for a 90-degree bank from 120 feet of track to just over 64 feet, thus meeting the space confinement requirements and, as a result, providing an even more exhilarating ride.

Another example: Similar situation but now we have a customer wanting to build a steel roller coaster in a very confined space. Maybe the ride is going to be located inside a compact building. Therefore, the curves are going to need to be very tight with small radii and will create large g-forces on the riders. The tight curve forces the vehicle length to be shorter to negotiate the course safely. To keep capacity up and waiting times down (because "I love waiting in line!" said no one ever), several vehicles will need to be linked together. A joint or coupling will need to be designed. Other criteria need to be deliberated: the coupling must allow each car to rotate relative to each other. The coupling must bear the loads during normal operation and should be easy for maintenance to inspect and, if necessary, be able to replace the component as quickly as possible.

One solution could involve using two articulating joint shafts, one inside of the other, separated by brass ball bearings. The outer shaft bears all the radial and axial loads acting on the joint during normal operation. The inner shaft does not take any load while the outer shaft is operable, allowing it to rotate freely. A lever is attached to this inner shaft so a maintenance technician can quickly inspect the mechanism. If the inner shaft can spin then the outer shaft is working fine, but if the inner shaft cannot spin then that means the first shaft has failed and is inoperable. For this reason, the inner shaft should be as strong as the outer shaft. During operation, if the outer shaft should fail the ride can continue safely, until maintenance can fix the outer shaft.

Figure 25 – Vekoma SLC wheels

These are just a few example scenarios and real-world issues that roller coaster engineers would be expected to solve! Sound like fun?

How a 4th Dimension Coaster Works

One of the popular breeds of roller coasters being built at a rapid pace today are called "wing" coasters (or "wing riders" or "4D" coasters). Wing coasters feature seats cantilevered off the side of the train instead of being on top of or below the rails allowing passenger's feet to dangle freely. There are currently four variations of this concept:

Type 1: Fully winged
Nickname: Wing Rider or wing coaster
Example: Wild Eagle at Dollywood

Type 2: Partially winged
Example: SkyRush at HersheyPark
There's a row of four seats and the outer most seats are to the sides of the track while the center two are still over the track.

Type 3: Fully winged with free spinning seats
Nickname: Intamin's ZacSpin, S&S 4D Free Spin
Example: Green Lantern at Six Flags Magic Mountain, Joker: The Ride found at various Six Flags theme parks.

When it comes to roller coasters where the seats spin to flip the passengers head-over-heels, the axis of rotation is very important. Have you ever watched a gymnast while she's doing a front flip? Where is the center of rotation at? The gymnast rotates somewhere around their midsection.

Green Lantern at Six Flags Magic Mountain was the first (and only) Intamin Zac Spin coaster built in the United States. It operated from 2011 to 2017 before being torn down – and for good reason. Green Lantern was very poorly received with many calling it rough and painful. The popular YouTube channel Defunctland has an entire episode called "The History of the Worst Six Flags Coaster, Green Lantern: First Flight." Why was the ride experience so bad?

The axis of rotation. On a Zac Spin riders sit back-to-back, and the axis of rotation is somewhere behind them, resulting in an uncomfortable experience. Another problem with the free spinning Zac Spins is the spinning is totally dependent on weight distribution: If there is a passenger on one side but not the other, the unbalanced load will result in some crazy flip-age.

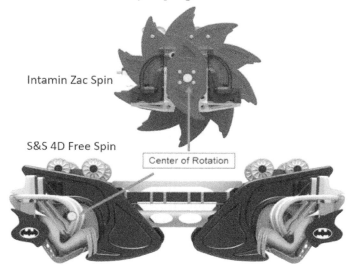

Figure 46 – Axis of rotation for different spinning models

Only four Zac Spin models were built worldwide before S&S Sansei came along and improved the design. Their 4D free spin coasters have a few improvements, most notably the positioning of the axis of rotation near rider's midsections resulting in more natural feeling flips. They also use Rocky Mountain Construction's Iron Horse track which is much more precise than a traditional steel coaster – something that takes on even more importance when you're creating wide trains with seats sticking out away from the center of gravity. The 4D free spin models are smoother as this type of track makes it easier to maintain plane resulting in no extra flapping or bouncing and less vibrations than the Zac Spins or previous 4D coasters.

Another area of improvement: the seats are free to spin but they also have a magnet and fin system that helps control the flipping. There are two sets of magnets: the first prevents the spinning from becoming too fast or too much while the second can be used to help induce spinning to flip the seats at the desired locations. As a result of an improved ride experience, twelve 4D free spins have already been built since 2015 and wouldn't be surprised to see more in the future.

Type 4: Controlled rotatable seats
Nickname: 4th dimension or 4D
Example: X2 at Six Flags Magic Mountain

The most complicated type of wing rider is the 4th dimension roller coaster, a highly unique variety of thrill machine in which the seats are cantilevered on each side of the vehicle (as opposed to being above or below the track), allowing the carriages to rotate 360 degrees. This controlled spinning or rotation is in a direction that is independent of the track – hence, it is like a fourth dimension. There are two sets of rails – one supports the weight of the vehicles while the other is what makes the seats rotate. The vertical distance or displacement between the two sets of rails controls the rotation of the passengers by transforming linear motion into rotational motion, accomplished via a rack and pinion gear.

Figure 47 – 4D coaster track showing how the second rails move

The pinion, a typical circular gear, engages the teeth on a linear gear bar (also known as a rack). Thus, as the spacing between the rails change, the wheels connected to the rack move vertically up or down, causing the pinion gear (or gears) to rotate, flipping the seats as much as 720 degrees. Pushing the rack up cause the seats to spin in one direction whereas pulling the rack down causes the seats to flip in the opposite direction. The amount of rotation is proportional to the displacement between the two sets of rails. No separate power supply is required; the forward motion of the vehicle due to gravity is enough. The pinion gear may use a complete gearbox to achieve the perfect ratio of linear to rotational motion. Of course, it's not as simple as it sounds because there must be flexibility built into the system due to vibrations and imperfections in the manufacture of the rails. Future versions may see the use of rotary hydraulic motors instead of applying the described arrangement.

How do the engineers know how to calculate the exact amount of rail movement required to obtain the desired rotation? Every pair of mating gears has a *gear ratio,* which is calculated by counting

Figure 48 – X2 at Six Flags Magic Mountain

the number of teeth on each gear and then dividing the number of teeth on the driven gear (or the output gear) by that of the driver gear (the input gear, that is, the one supplied the power). For example, a gear with 30 teeth drives a gear with 90 teeth. Dividing 90 by 30 gives a ratio of 3/1, meaning that for every three rotations the driver gear makes, the larger gear (which is the driven gear) turns once. It sounds a little confusing at first but is really quite simple!

A rack-and-pinion gear system consists of a round gear known as the pinion and a flat, toothed component known as the rack. The principle is the same. However, rather than rotations, the ratio determines the linear distance traveled by the rack with each rotation of the pinion. In this case, instead of counting the number of teeth in each gear, you measure the distance moved by the rack in inches. To get the gear ratio, all you must do is measure the distance from the end of the rack to an arbitrary point; turn the pinion one full revolution and then measure the distance again, with the difference being the gear ratio.

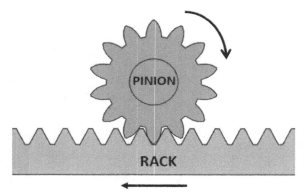

Figure 49 – A simple rack and pinion gear

Only three 4th dimension coasters have been built worldwide, the most recent in 2012. But 4D coasters are not the only ones to use a rack and pinion...

A Roller Coaster on a Cruise Ship

At the 2016 IAAPA Expo in Orlando, one of the attendees showed a very different type of roller coaster, the Spike Coaster. Spike is a new type of powered coaster developed by Maurer Rides, and the first prototype was tested in Germany. The Spike Coaster is a powered coaster that is driven by a rack and pinion, like in a car's steering system. The "rack" is a row of gear teeth that is fixed along the length of the track also known as the "counter-toothing"), while the "pinion" is a gear on the ride vehicle that turns to propel the car down the track.

Rack and pinions have been used for transportation since the 19th century in "Rack Railways," but the key development for Maurer was the ability to make the rack into tight, compound curves. This allows the Spike to follow the high-speed designs of a roller coaster track. The technology was developed primarily for the roller coaster field, but in parallel Spike's "people mover" design was developed, which is primarily being marketed for places like ski resorts where a rack railway that can operate in all weather conditions could be useful. Spike views the technology as usable in a wide range of transport fields, but primarily for roller coasters.

Figure 50 – See the rack or "spike" attached to the rail?

As mentioned, Spike is a rack-and-pinion coaster, driven by a motor on the ride vehicle. The other obvious form of powering a vehicle along the track is to use drive tires, but the rack and pinion offers several advantages over drive tires. The biggest benefit is that the gears allow for faster accelerations and speeds through turns. Drive tires can slip if the vehicle accelerates too much, just like a car that skids. On turns, the tires on the inside and outside of the bend get squished a different amount and must move at different speeds (as the outer tire goes through a bigger radius), which also slows the vehicle. The rack-and-pinion eliminates these problems.

One of the downsides of the Spike powered coaster design is that because it's propelled by a drive gear, it doesn't actually "coast". That means the sensation on drops and airtime on hills won't necessarily be like a normal coaster and might not have the same thrill. However, one way Maurer can make up for this is by doing what they call "Compressed Airtime." Because Spike is driven, it can accelerate FASTER than gravity when going down. This allows Spike to create airtime on sections of the track that wouldn't normally have that sensation. It also means that designs aren't beholden to the amount of kinetic energy from the initial hill or launch. This allows turns that can be made five meters vertical and three meters horizontal in any layout you can dream up with no energy calculation needed.

The way the Spike coaster is designed, the rack teeth on the track won't wear away over time, so it should be low maintenance. This is achieved by using polyurethane drive gears on the cars. The drive gears can be easily changed in the same way wheels are changed on normal roller coaster cars. Thanks to the fact that the polyurethane will wear preferentially compared to the metal teeth bolted to the track, only the wheels will need regular replacement, keeping maintenance costs down.

One of the other great benefits of the Spike design is that it can both be pre-programmed to go at certain speeds throughout the track (like in a way that, say, Radiator Springs Racers at Disney's California Adventure works), or the individual vehicles can be controlled by riders. The vehicles would have certain minimum and maximum speeds throughout the course, but you could design in certain "boost zones" where guests could hit the accelerator. The vehicle's speed limit can also be different depending on the area of the track to ensure the rider's safety or change the experience. Designers can say, "this section is only going to draw X amount of power maximum" and even if a rider is using the accelerator to go full blast, the power draw and therefore the speed is limited until they leave that section.

The flexibility with speeds means that you can also use the same track for different levels of thrill. Different track profiles can be created on the same track. One hour can be aggressive, the next hour could be for smaller kids. Or every ride can be different. It also means a ride could be updated over time so that each year it's a "new" ride.

As for how the braking and block sectioning works, it's sort of the opposite of a normal coaster. There are no sensors on the track, all sensing is done by the vehicles, which report it to a main controller via Wi-Fi or an extra "communications" rail. The vehicle position is determined by counting the gear teeth, like how an odometer on a car works by counting wheel rotations. Each vehicle has its own block zone, like a bubble around it, that other vehicles can't enter. So, if one vehicle is going slowly and another catches up to it, the trailing vehicle will have its max speed limit reduced to keep it out of the "block zone" of the slower vehicle, and the slower front vehicle may be sped up.

This blocking is a big part of what allows the capacity to be kept relatively high even though there are single vehicles rather than long trains. Maurer claims they can get dispatches up to as quickly as every five seconds, which can give capacities up to 1,400 riders per hour, depending on the track length and speeds. However, if there are more vehicles on the track at once, a guest has less room to play with speeding up or slowing down a vehicle if it isn't pre-programmed, so it is a balance for the designers.

Spike currently offers several different vehicles that can be used on the same track, some with riders on the outside (like a wing coaster) and some with riders on top. From one of their patents, this is possible because "The center of gravity of the laden or unladen vehicle is always above, albeit as close as possible to, the first and/or second guide element. Thus, a seat arrangement can be provided, wherein at least one of the rails (first and/or second guide element) is arranged between the legs of a passenger or at least one of the rails is arranged between two adjacent seats." This variety also means that a ride can be updated with different vehicles to offer a new sensation.

As mentioned before, each individual vehicle is powered, which also means that individual vehicles can have effects. The obvious is things like lights and decorations, but also the potential for things like an interactive digital display. The individual powered vehicles also lead to lower power demands on the ride thanks to the fact that the vehicles are small and lightweight compared to a long train. Because each vehicle launches individually, the peak power requirements are much lower. Regenerative braking on the vehicles also helps reduce energy requirements.

Currently the vehicles are the primary limiter on how extreme certain aspects of the Spike can be. The rides all follow safety standards from ASTM for amusement rides on how much the body can be pushed, but aspects of the vehicles like the restrains are what limit acceleration and don't currently allow inversions in the designs. Future vehicle designs could remove some of these limits. One of the other current downsides is that the individually powered vehicles mean a disabled vehicle is stuck on the track. The vehicles can be towed or pushed by other cars to a safe evacuation area, but Maurer is currently working on a way to evacuate riders wherever a stranded vehicle may be.

The first Spike coaster to open was the Sky Dragster at Skyline Park in Bavaria, Germany in 2017. The layout was short; 889 feet of track cramped into a 65.6 ft by 328-foot rectangle. The next Spike coaster to open was much larger, the twin 1,722 foot long tracked Desmo Race Mirabilandia in Italy in 2019. But what really grabs your attention is what came next.

Like theme parks, the cruise industry is super competitive with bigger and better ships always looking for that "next" innovation. Carnival Cruise Line's Mardi Gras debuted an industry first when it set sail in 2021: the first-ever roller coaster at sea. Built by Munich-based Maurer Rides, BOLT: Ultimate Sea Coaster is a heart-pounding rush of adrenaline offering nearly 800 feet of exhilarating twists, turns and drops with riders reaching speeds of nearly 40 miles per hour with gravitational forces of 1.2 g. BOLT is an all-electric SPIKE roller coaster that allows two riders in a motorcycle-like vehicle to race along a track 187 feet above sea level, enabling guests to experience the sea in an exciting new way with breathtaking 360-degree views.

The coaster begins with an action-packed launch where riders can achieve race car-like levels of acceleration and culminates with a high-powered hairpin turn around Carnival's iconic funnel. Riders' speeds are posted after the race, and just like land-based roller coasters, guests have their photo taken during the ride for a memorable keepsake. Guests will be able to choose their own speed, making each ride unique.

It's no simple task to add a roller coaster to a cruise ship. The complexity of it is probably the main reason it hasn't been done before. Besides all the normal challenges of designing a roller coaster on solid land, now you must deal with the additional challenges of a moving, rocking ship, corrosive salt water, potentially high winds, as well as space and weight limitations. Adding a ride as complex as a roller coaster needs to be developed early in the processes of designing the ship – you can't just tack it on later. The development of Carnival's new class of ship coinciding with Maurer's development of the Spike Coaster system worked out to be perfect timing.

Figure 51 – The twisting black track of Bolt under construction

Mardi Gras' BOLT roller coaster does not rely on gravity to get around the track and it is an extremely quiet ride. These were two key considerations when selecting this ride system. Traditional roller coasters tend to be very noisy. This wouldn't work on a cruise ship where guests are also relaxing and enjoying other activities in the area. The Spike Coaster system is so quiet that guests will barely hear it. It is also a powered ride all the way through, and the vehicle has constant traction with the track making it a perfect fit for a roller coaster experience on a moving ship.

Bolt was completely assembled on the ground in Germany before it was attached to the ship when it was berthed in Turkey. Some of the structural elements, like supporting columns, are quite similar in appearance to other attractions onboard the ship, the difference comes in the engineering and design to account for the movement of the vehicle around the track as well as the movement of the ship itself. One of the many advantages of this feature is that the 722 feet of track is suspended above the deck of the ship leaving a lot of space below available for the many other exciting amenities and features. The planning and development of this coaster is different because it had never been done before, so it required extreme attention to detail and close collaboration with the shipyard, Maurer's engineering team, experts in safety, noise, and vibration. More Spike coasters are expected to hit the seas soon.

How a Roller Coaster Jumps the Tracks

You've all seen those YouTube video thumbnails before. You know the ones I'm talking about. Where there's a picture of a roller coaster's loop and half of the track has been removed via photoshop, so it looks like the coaster cars are jumping the tracks. And don't forget about the big red arrow and click-bait title. I hate to spoil your dreams, but no real-world roller coasters actually completely leave the track(yet).

However, Universal Creative is producing Donkey Kong-themed coasters for several of their theme parks throughout the world that will create the illusion of the cars jumping the track. Here's how:

A mine cart themed ride vehicle will look as if it is attached to a railroad type track system, but in reality, the car is attached to a boom arm that goes down under the cart and attaches to a typical roller coaster track entirely hidden from view by themed elements. The concept would allow for the hidden coaster track to go up and down hills without the riders being able to see what is coming next. For example, it would allow for the mine cart to appear to "jump" a gap in the scenic mine cart track, while staying connected to the hidden coaster track system below.

One challenge for the designers of the boom coaster is the rider's heartline being so far away from the coaster's track. When banking into a curve, if the axis of rotation is about the spine of the track, then the center of gravity of the passenger is accelerated toward the center of the curve. This results in the passenger being thrown against the side of the car or against another passenger. Roller coasters must not put too much strain on a rider. To correct for this acceleration, the track should be rotated around the center line, or heartline of the passengers (roughly just above the center of a human torso). Thus, the acceleration of the passengers to the inside of the curve is greatly reduced. The track would have to swing out wide when going into a banked turn. Can't wait to see these coasters operate in the coming years.

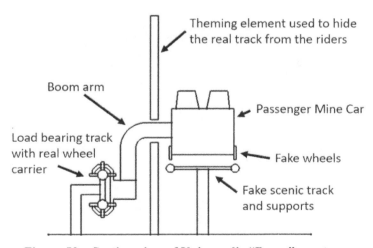

Figure 52 – Section view of Universal's "Boom" coaster

Chapter 6: Infrastructure

Supports and Foundations

One of the most important, and perhaps most overlooked, components of a roller coaster are the foundations. It's an exciting time when you go to your nearest amusement park and see holes in the ground. This is a sign that foundations, more commonly referred to as "footers," are being dug and a new ride is about to be installed. Footers must support the weight of the support columns, track, and fully loaded trains. The ride must also be built to withstand earthquakes, soil erosion, tornados, floods, or other environmental hazards depending on its location. The maximum static and dynamic design loads of each footing or equivalent structural connection must be defined. The size and shape of a footer depend on the loads and soil conditions.

Figure 53 – Footers for a new roller coaster

The main engineering principles to understand here are tension and compression. When a specimen of material is loaded in such a way that it extends the material then it is said to be in tension. Conversely, if the material shortens under axially directed pushing forces it is said to be in compression. The capacity to resist such forces is called compressive strength. When the limit of compressive strength is reached, materials are crushed.

When a coaster train is rolling down the bottom of a hill it pushes the track downward causing a compressive force on the foundation of the structure due to the weight of the train and the positive gravitational forces acting on the ride. When a coaster train flies over the top of a small hill that provides airtime and even negative g's the wheels on the underside of the train want to pull the track upward. This upward "stretching" of the structure is a tension force. If these forces are too great, they may cause the foundation to crack.

Figure 34 – Support

What does this have to do with footers? The main material in the foundation is concrete, which can have a very high compressive strength. While concrete is strong in compression, it is weak in tension. However, steel is strong under forces of tension, so combining the two elements results in the creation of a very strong foundation. This is called reinforced concrete. Steel is embedded inside the concrete in such a manner that the two materials act together in resisting forces.

Compressive force is transmitted to the foundation through perfect contact between the column base plate and the grout poured between the foundation and the support column to fill the gap. What is grouting and why should we care? Imagine standing barefoot on top of six bolts all day, everyday… now imagine standing on top of six bolts but this time being supported by the world's best shoe!

Grouting can be described as the shoe of the coaster allowing the force to be quickly and effectively dispersed down the column, through the grout, then through the foundation.

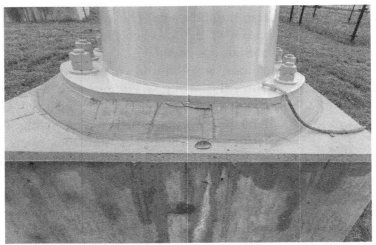

Figure 55 – Example of grout between steel column and concrete footer

When assembling support columns during construction, removable pins are used to align the bolt holes in the two columns before inserting the larger bolts and tightening the nuts. The column is first set and put into position with the adjustment gap (the empty space between the column and the foundation). Once the column and track are precisely aligned and locked down, this extra layer of concrete known as grout is applied to offer full resting support for the coaster.

Steel bolts embedded within the concrete hold the column to the foundation. Shear force is transmitted to the foundation through perfect contact between the shear pin and grout as well as between the anchor bolts and grout. Tension force is transmitted to the foundation through the anchor bolts.

Another type of foundation is to have a horizontal steel structure laid on the ground, which connects multiple supports into a lattice. This method is often used in traveling coasters as well as locations that don't allow pouring concrete foundations like piers.

This type of construction allowed the Columbus Zoo and Aquarium to tear down a log flume and open their Zamperla spinning coaster Tidal Twist in just 97 days!

Figure 56 – Example of a steel lattice foundation

The structural interface for the ride needs to be analyzed for safety assurances. Every single connection holding every component of the roller coaster together will be closely examined. Maximum static and dynamic load distribution for each concrete footing need to be specified. Every fastener needs a defined grade, torque, and replacement or tightening schedule. All these components working together help keep a roller coaster standing upright.

In general, roller coasters shall be designed so the expected loading conditions will not cause stress to exceed the yield strength of the materials. That means no significant plastic deformation should occur when structures and components are subjected to expected loads (with the exception of earthquakes).

Brakes, Blocks, Sensors, and Switches

High performance roller coasters today can reach speeds more than one hundred miles per hour, yet still slow down safely and efficiently, returning their passengers to the unloading station unharmed. Control is the ability to stop the ride. Yet, there are no brake systems attached to the vehicles themselves, meaning the vehicle can only be slowed in locations where there are brakes fixed to the track. Roller coasters primarily rely on two types of brakes: friction and magnetic.

Friction brakes consist of two opposite shoe clamps that close on a central beam or fin mounted to the underside of the vehicle's chassis. The clamping force is given by two springs located inside two pneumatic jacks. The brakes default position is closed for safety reasons and the fins are only able to open to allow the car to pass by applying air pressure. The downside to using friction brakes is the amount of heat created and the wear and tear on the components. They can also be affected by humidity or moisture.

Figure 57 – Example of friction brakes with sensors

Scores of modern mega coasters are now making use of magnetic braking systems. This technique involves mounting permanent magnets on the track to oppose the motion of the cars traveling past by simple magnetic repulsion. It's kind of magical to see a fully loaded train slow down without touching anything. Because there is no longer physical contact between the vehicle and the brake, the system is much easier to maintain and much quieter. Magnets do degrade over time and need to eventually be replaced, but not as frequently as friction brakes.

Additionally, because the system imparts a force based on the velocity of the train (as opposed to a harsh static frictional force), the deceleration is smoother. With the help of pneumatics, the assembly can rotate away from the track and completely remove themselves from affecting the ride's speed, such as on a launched coaster track where the brakes disengage to allow a vehicle to pass but then reengage in case of a rollback (when the train doesn't make it over a hill and rolls back). Of course, whenever you make a component move mechanically there is another potential point of breakdown that might require repair and cause downtime.

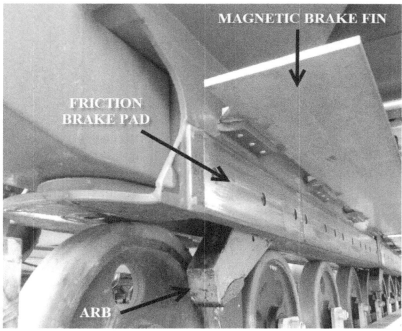

Figure 58 – Brake components on the underside of a vehicle

However, permanent magnetic brakes cannot completely stop a train as the magnitude of the force is proportional to the velocity. The higher the velocity is, the stronger the braking force will be. Thus, why most coasters use a combination of magnetic and friction-based braking mechanisms working in tandem to provide safe rides.

There is a third, though far less common, type of brake: the water brake. The car plunges into a pool of water or the trains have "scoop" attachments that contact the water to trim off some speed. Water level must be tightly controlled – not enough and the car will not lose enough speed. Too high and the impact with the water could turn dangerous.

Figure 59 – Splashdown finale on Diamondback

To have the highest throughput possible, modern roller coasters operate more than one train at a time. A sophisticated monitoring and control system is required to run multiple trains on the same track safely and efficiently. This protective system, which prevents any two trains from getting too close to one another, is called a "block system" and is composed of a series of "blocks." A block is a section of track with a controllable start and stop point. Two trains should never ever occupy the same block at the same time. If a train were to not clear a block for whatever reason, the following train needs to be stopped before it has a chance to plow into the back of the unsuspecting passengers.

No train is allowed to enter a block until the train that preceded it has safely exited the block. Therefore, each block must contain a method for completely stopping a train and a way to get it moving again. A train can be stopped by mechanical brakes or by disabling the device which moves the vehicle such as a kicker tire or a chain lift. Methods to propel the train from its dead stop can be accomplished by re-starting the chain lift, using drive tire motors, or using good ole gravity by designing the area of track where the train stops at a downward sloping angle (equal or greater than the neutral slope). Wherever a train can be stopped on purpose, a safe method of evacuation should be provided.

Figure 60 – Drive tire motors

A roller coaster must have at least one more block than it does trains. A typical coaster might have the following block sections which can stop the train if needed:

1. Station
2. Lift Hill
3. Mid-Course Brake Run (MCBR)
4. End Brake
5. Transfer Track

Every block section must have a method to completely stop the train but not every brake section is considered a block. Trim brakes are used to reduce the speed of a train to homogenize every ride to fit comfortably within the desired speed envelope, meaning every ride feels relatively the same regardless of the operating conditions that affect the speed variances. Trim brakes are often found on the way up airtime hills so they can help prevent pulling the structure up out of the ground, so it doesn't fatigue the structure. Trims are sometimes added after the fact, as reducing speed can prevent excess wear and tear as long as the cars can still have enough energy to make it back to the station. The ride computer monitors the speed of the train with sensors before the trim and engages or disengages the trim brake devices to reach the required exiting speed at the end of the trim brake. How?

Figure 61 – A trim brake on Raging Bull

The entire blocking system relies on a sophisticated computer or Programmable Logic Controller (PLC). The PLC is a special-purpose computer which simply takes real-world signals, makes decisions based on those signals, and produces output signals based on those decisions. For instance, a PLC can easily be equipped with a switch to count lap bar release pedals so that it can detect if a restraint is left open. It can be connected to an anemometer to measure the wind speed and set brake pressures accordingly. The PLC can even detect train speed at various points on the course and monitor the ride's performance.

These days, nearly all roller coasters use PLCs to control the block system and the rules remain the same as they always have: No more than one train may occupy any block on the ride at any time. To do this the computer must know where the trains are at all times. Trains are detected in a variety of ways:

Proximity (proxy) switches are the most popular method used to detect the position of the trains on the coaster's track. These are electromagnetic devices located at intervals along the track which are programmed to detect the presence of a metal object. The individual cars of the train will have a metal component attached to their underside, often called a flag. Using the proximity switch, the computer can count the number of cars that go by and signal whether the entire train has safely passed through the block depending on the number of flags it counted. This type of system leaves little room for error which is why it is widely employed.

Another method used on roller coasters to detect train position is called a photo eye. Photo eye detectors are versatile in the fact that there are several different setup configurations. A single eye can bounce a beam of light off a fixed reflector mounted on the support structure. When the train passes the eye, it breaks the beam of light signaling to the computer the location of the train. Instead of using a fixed reflector a pair of photo eyes can be built directly across from each other on either side of the track. A third method is to mount the reflector onto the side of the train so that when it passes it bounces the light back to the photo eye. This method is least popular because a mechanic, carpenter, or film crew could accidentally break the beam and stop the ride by confusing the control system.

Another, less popular, method of detecting a train is a mechanical device called a "limit switch" where a moving arm mounted on the track is designed to come into contact with a passing train. This limit switch is spring loaded causing the arm to be pushed forward as a train passes overhead. After the train has passed the switch arm returns to its home position. These types of switches are more prone to malfunctions due to their mechanical nature.

Gravity coasters with few places for the train to stop do not have a large number of sensors. A simple five-zone

Figure 62 – A switch used to detect position of the train

system (like described earlier) with only two trains or vehicles could only have a couple dozen sensors. In comparison, a twelve-vehicle system may have hundreds of sensors to provide the necessary granularity of sensing, particularly in areas such as the brake run and station where vehicles are close together. The number of sensors required increases with rising capacity and lower dispatch intervals. Sensors are not only used to detect the position of the train, but they can also detect the speed or activate scenic elements situated along the track.

There is also the option today for a coaster to use a "rolling block" system. On a traditional coaster the station is one block zone and the first train must fully leave the station before the second one enters. A rolling block allows the second train to begin entering the station at the same time as the first train is leaving it, thus saving a few precious seconds, and increasing efficiency.

Racing/Dueling Control Systems

Dueling roller coasters offer a thrilling element not typically found on single track rides: the near-miss collision. Two trains travel directly toward one another often at a combined speed more than one-hundred miles per hour, seemingly on a head-on collision course, only to veer away at the last breathtaking second, narrowly avoiding the other train by a few inches. But how do two trains on two separate tracks arrive at the same point hundreds or thousands of feet into their layout time after time?

Without perfect synchronization these near-miss collision elements are useless and defeat the point of building an expensive dual ride if they are not timed up exactly. After the moment of release from the highest point on the ride there is no way to control the speed of the trains until the end (or a mid-course brake run). Dueling coasters work by weighing the trains and then by using past performance data to determine which train to release from the lift first and how much of a head start to give it.

On a regular, single track roller coaster, small differences in speed and duration of the ride don't matter too much as long as they're still within the operating envelope. However, they can greatly affect the ride experience on a racer or dueler. Near misses can only be achieved if the vehicles are exactly the same. In the real world, this is impossible. Each train will have a different weight due to number of passengers; some wheels will be more worn than others, and so on. So, what do you do to consistently ensure perfect synchronization? The engineers design the ride where the variables such as aerodynamics and rolling resistance can be controlled as much as possible.

The main variable to compensate for is weight. But how do you weigh a fully loaded coaster train? The solution is pretty simple. The weight of the trains can be determined by measuring the current draw on the lift hill motors (or LIMs in the case of a launched ride). This can be done because the weight of the loaded train is directly proportional to the power needed to pull the train up the lift hill. For example, say you know the weight of an empty train. You can measure the current draw on the motor as the train proceeds up the lift. Next, add a few water dummies with a known weight to the train and take another current measurement. Now we can interpolate between those numbers to determine other unknown weights using the current draw alone.

The next issue is: how do you know how much of a delay to release the second train relative to the first? The goal is for the second, faster train to catch up to the lighter, slower train at the desired location. Several factors which influence the speed of each train include aerodynamics (surface area), rolling resistance, wheel condition (worn), track condition (greased or not), condition of tires, bearing condition, wheel alignment, etc. All these factors can cause a faster or slower than expected ride and the tricky part is that they are different for every circuit of the course. Wheel wear may not be an issue in the morning but by the end of the day, after hundreds of circuits, the train may be performing much differently than just a few hours earlier.

This problem is solved by graphing a train performance curve which will constantly be updated. Any change in the tire's rolling resistance or vehicle aerodynamics is instantly compensated for. In theory, the near-miss collisions should get better and more closely timed as the day goes on and as more data points are added to the performance curve. The computer can take real time data and combine it with past performance results. Similar performance curves may be implemented on launch coasters. That data can help determine how much power needs to be applied to the motor to launch the train over the hill.

Dueling coasters contain multiple near-miss elements but there is often no way to adjust the speed or timing of the trains after they are released from the lift hill. Therefore, every near miss cannot be controlled exactly. In this case, the designers must pick which element they want to have the best duel. Other coasters have the additional challenge of one train being floorless while the other is inverted.

As the trains are pulled up the lift, the control system measures the weight of each train by determining the amount of current drawn on the motor. Sensors tell the computers where the trains are on the lift. The computer calculates which train to release first based on a performance curve of data which is updated after every circuit. The lift hill motors are adjusted accordingly. Dueling roller coasters have increased the thrill and excitement level by using a control system which allows two trains on separate tracks to simultaneously arrive at the same near-miss location repeatedly.

Safety Systems on the World's Scariest Roller Coaster

A ride on Gravity Max at Taiwan's Lihpao Land theme park begins innocently enough with a typical chain lift hill. But upon cresting the apex, you'll notice the track suddenly ends – the rails stop, and you see nothing but blue sky. The train keeps creeping forward until it reaches the end of the track, thankfully coming to a stop.

What happens next will blow your mind—the entire section of track the vehicle is now sitting rotates ninety degrees. You're on the world's first tilt coaster and in a terrifying heartbeat you're staring straight down at the ground. The movable track segment is locked into the next section of track, and, without warning, the train is released causing you to free fall over one hundred feet at fifty-six miles per hour. Scream your head off but rest assured there's no reason to be scared for your safety on one of the world's scariest roller coasters.

Vekoma tilt coasters have redundant safety features to prevent the train from leaving the tilt track prematurely. A hook at the end of the track latches onto the back of the train. There's also a block at the front of the track that would prevent the train from smashing into the ground if the hook were ever to fail. When the coaster tilts to ninety degrees, there are several sensors to check the position of the track to ensure the rails are aligned before the computer gives the OK to release the train. Occasionally, the rails will fail to align, usually due to a weather-related event like rain or wind. When the rails can't fully align the computer doesn't release the train. Instead, it lowers the tilt track back to the horizontal position and all twenty riders can be safely released from the train and onto the horizontal platform designed for exactly this purpose.

Gravity Max is no longer the only tilt coaster in the world. Golden Horse has four "broken rail" coasters open in China including the dueling tilt coaster The Battle of Jungle King. Escape from Gringotts at Universal Studios Florida is also a tilt coaster with it's see-saw track element, it just doesn't go all the way to ninety-degrees. Austin, Texas' Circuit of the Americas (COTA), known primarily for racing, is about to pop up on the radar of more roller coaster enthusiasts (or at least your crazy relatives who tag you in Facebook posts asking, "Would you ride this?!?!", COTALand will be home to North America's first new age Vekoma tilt coaster, Circuit Breaker, which is scheduled to open in 2023.

Chapter 7: From Paper to the Park

Manufacturing

Steel coaster track typically consists of two pipes, or rails, where the wheels ride. The rails are held at a constant pitch by steel pieces called ties. Sometimes there is a third pipe in the middle of the track called a spine. Saddles connect the track (through the spine) to the support columns. Sections of track are welded together in a factory before being shipped to the construction site where they are bolted together. Different manufacturers have different track designs.

The length of each track segment is dependent on the method of shipping: flatbed trucks or shipping containers if traveling by boat. The longer the track section lengths can be, the fewer joints, and a smoother ride. A single piece of steel coaster track may weigh 30,000 pounds. Steel track pieces are joined together by bolts. When bolts are torqued properly, they stretch slightly and behave like really strong springs, holding components together rigidly, so that they behave as a single unit. If the bolt is loose, movement of the bolted parts can apply impact loads that can cause the bolt to fail.

Steel coaster track and support design is all about efficiently using materials to help keep costs down. Engineers use software to minimize weight and the amount of material needed while maximizing the work the material can do. The thickness of the tubes is designed to be exactly what it needs to be for each location on the ride. High stress areas like the bottom of the drops will use thicker steel pieces than low stress areas like the station track. It's cheaper to produce simple support columns rather than bending steel pipes, so in low to the ground areas you may see more supports closer together rather than using a spine between the rails. But in tall coasters it makes sense to use fewer supports, so the size of the spine is increased to allow for greater spacing between the columns. The more track the higher off the ground, the more material required, the more expensive the coaster will be.

Typical steel coaster tubular track is made from hot rolled mild steel. Other steel components are galvanized by submerging the bare steel into molten zinc. The iron in the steel reacts with the molten zinc to form an alloy coating, thus giving it incredibly high corrosion protection. This protection is important to keep the ride smooth, reliable, and thrilling for many years.

One drawback of using bent steel rods or pipes is the resulting shape is not always one-hundred percent accurate, a potentially large issue when you are dealing with thousands of feet of pipe. Steel will typically seek to bend at its weakest point or where the most force is applied over a span. Manufacturers then attempt to fix the pipe by bending it again or end up settling for a less-than-perfect piece.

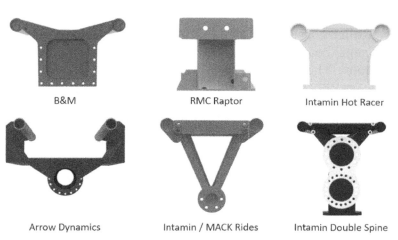

Figure 63 – Manufacturers have different track types (not to scale)

New manufacturing techniques are being invented to minimize manufacturing stresses, increase accuracy, and reduce cost. One example of a relatively new approach is where the rails are fabricated by welding together custom cut planar pieces of steel with minimal heating or bending required. The top rail piece still needs to be bent but it is a lot easier to bend a planar sheet of steel compared to a hollow tube. Of course, given a big enough stock piece of steel one could machine out the exact shape needed, but that would be very impractical (time consuming with lots of waste).

Special jigs and figures must be designed to hold all the pieces in place as they are joined together. Track segments are pre-fit together in the shop to prevent problems in the field during install. If the track needs grinding or welding, it is easier to do it in the factory where the environment can be controlled and is much easier to reach the required quality of welds.

After the steel pieces are welded together, the track is sandblasted and then painted. The sandblasting not only removes dirt and grime but leaves behind minuscule scratches that make it easier for coatings to sink into and bond with. This gives a uniform, smooth, appearance to the newly coated material while also ensuring that the coating will last as long as possible without peeling or cracking.

Roller coasters are painted in a high performance industrial polysiloxane coating system designed for industrial and offshore applications. The coating helps protect the steel against rust. The park always chooses what color they want for the track. Sometimes the manufacturer will recommend certain colors as some colors require fewer coats, vary in cost, and fade less in the sun. But ultimately the park chooses the colors they want. Over time the ultraviolet light in the sun's rays fade paint, making areas duller than the designers intended. Some colors are naturally resistant to fading, UV resistant varnish and two pack paint, are all part of the defense against fading.

Completed sections of track are then shipped to the construction site for installation, where they are bolted to the support structure and to other track segments. When a new ride is installed, it often comes straight from the manufacturer with a brand-new paint job. But the installation process can be harsh, resulting in burnishes and scrapes or scratches that always need to be touched up throughout the ride. Then you have thousands of fasteners, literally the nuts and bolts that hold the thing together, that will be in need of paint once the installers are finished. Painters usually spend a week or two on brand new coasters making sure everything looks its best after installation is complete, and just before the ride is opened to the public.

Construction

Put on your civil engineering hat because it's time to begin assembling the roller coaster on site. Before construction begins a location must be chosen where parts can be stored. If building a wooden coaster or a structure that needs pre-assembly on the ground before being erected, then there must be a location to do that. The project manager must decide whether to build structure in pieces and carry it via forklift to the construction site but there could be extra expenses in transporting materials. Additional factors that complicate construction is if the park is open or closed and are there any operating rides to build around. The longer it takes to build a ride, the longer the customer must pay for the construction staff and equipment. It's up to the project or construction management to build the ride efficiently.

One of the first steps in construction is to have surveyors plot out the course of the coaster. A predetermined reference point is set, and precise measurements are taken. The soil underneath the centerline of the coaster will be tested to establish its condition. Certain areas, such as marshes or sand bars, are so poor that pilings need to be driven deep into the earth to find solid bedrock to support the weight of the coaster. The concrete foundations for the ride are designed specifically for the soil conditions. It is typically the responsibility of the park to provide the geotechnical data to the designer at the onset of the project. Even with this data, surprises still happen, such as when you're digging a tunnel and hit a previously unmarked underground spring.

Before proper construction can begin, the ground must first be prepared and leveled off to the correct elevation. Here, heavy construction equipment is used to grade the ground according to the blueprints and remove any trees or foliage. Only necessary trees are removed to keep as much of a forest atmosphere as possible. Some obstacles cannot be removed, such as any existing buildings, plumbing, power lines, or other infrastructure. The surveyors use trigonometry to plot around these objects.

Once the ground preparation is completed, wooden reference markers are added to the site. These stakes, once placed, should not be moved. Nails are then hammered in precise locations on the stakes so that measuring devices can be attached. This is performed to help find the intended center of a hole where a foundation (also known as a footer or pier) is going to be poured. A common measuring device is the plum bob, a pointed dagger like instrument on a string. Once the location of the footer is determined, steel reinforcement bars are submerged in concrete and allowed to sit for the specified duration of time it takes to harden before the support columns are installed.

Often roller coasters require the engineers to also design large bridge spans to cross rivers or lakes. With the immense weight of the bridges, ride sinkage can come into play. Deep pilings need to be drilled and poured, raising overage cost of the project. To manage construction and engineering costs, load limits on bridges must always be reduced as much as possible to minimize bracing and steel required for the bridge. To keep the designs simple, engineers don't like putting horizontal loads in the steel, so bridges tend to be straight sections of track.

Once the steel support columns are erected, they are often filled with sand to reduce the noise levels of the coaster. Sand isn't the only stuff you can use. Expanding foam works just as well and is considerably lighter, although it is a bit more expensive.

Wooden coaster construction, on the other hand, is tedious, back-breaking work. Twenty-foot-long boards must be hauled up and down hills through the coldest, most miserable time of the year. Carpenters drive galvanized steel nails into rock-hard lumber or drill thousands of holes while hanging vertically on the side of the structure. It's not surprising to see there's always a chance the carpenter cutting the wood at the construction site might deviate from the design specifications by accident.

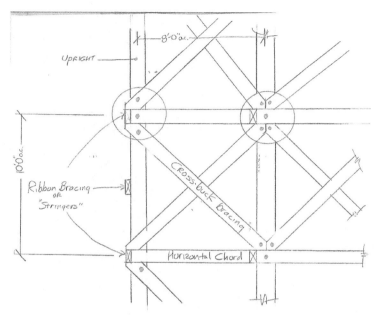

Figure 44 – Sketch of a wood coaster's structure

Wood coaster design may look completely random, but it is very precise, with each and every piece of board having its own name and purpose. With today's software, every piece of wood can be analyzed. Each piece has a use factor, or percentage of capacity. A wooden board may be stressed to 70% of its capacity.

Huge posts paired together and called bents are assembled on the ground then lifted into the air and placed on the concrete footers. Horizontal chords and diagonal braces are nailed between the bents to brace them. Ledgers, the massive boards at the top of each bent, is what the track will rest on. Walkboards are laid on either side of the track and give the carpenters, mechanics, and inspectors a solid structure to walk on while performing their jobs.

Typical wooden coaster track has rail stacks made from laminated pressure-treated hard-to-break southern yellow pine. The wood is placed in a vacuum chamber, where all the air is sucked out and moisture is removed. Next, to prevent decay, fungal growth, and insect infestation, alkaline copper quaternary and chromated copper arsenate are pumped in under pressure to force them into the wood. A thin strip of steel goes atop the completed rail stack to form a riding surface for the coaster wheels. Each wooden layer is a 2 × 10 or 2 × 12. The structure and rails have to support a fully loaded train accelerating at up to 3 gs.

Different companies have experimented with various materials to try to prolong the life of the wood coaster and reduce the amount of maintenance needed. Instead of using treated pine wood, certain parks have begun using a Brazilian hardwood known as "ipe" (ee-pay). Ipe is one of the hardest woods in the world which lends to its incredible durability and longevity. The super-strength wood has the same structural characteristics as steel — for instance, it's so dense that it doesn't float, and it has the same fire-resistance rating as steel. You can't drive a nail into it — all the holes must be pre-drilled. This stronger material prevents the steel track from digging into the wood, which is what necessitates a lot of wood coaster-related maintenance and rehab work.

The park must decide if they want to paint their wood coaster or not. Wooden coasters often use the same high quality acrylic paints you can buy for your own house, these days they have excellent warranties and hold up well to the elements. The main difference is they get the product directly from the manufacturer, with the colors ground in at the factory, instead of using the tinting systems most people are used to. This helps keep the colors perfectly consistent throughout the huge rides.

Roller coasters are built in phases. While the construction crew is erecting one section of the ride, the engineers could be finalizing the design and details for another. At the same time, the designers are also responsible for providing layout support and construction drawings to the crew, and support to the park for ordering and tracking material and equipment, and maybe even supplying the park's marketing and public relations (PR) departments with renderings and videos.

The construction of a roller coaster is a giant jigsaw puzzle the project managers must oversee to ensure everything comes to together in the right place at the right times. Kentucky Kingdom is located right next to the Louisville Muhammad Ali International Airport. When they construct a big ride, like the Storm Chaser RMC coaster, they must work with the airport to receive permission when they can fly their cranes to not interfere with any aircrafts.

With all the heavy equipment flying around, mistakes do happen. A piece of machinery fell on the Eagles Life in the Fast Lane Vekoma mine train coaster during construction of Hard Rock Park in Myrtle Beach (now defunct). The track was damaged and bent out of shape and had to be repaired.

As the physical structure with all the mechanical components is nearing completion, the electrical engineers arrive on-site to begin installing all the wiring and electrical systems according to their specifications. Control engineers produce all the drawings necessary for the systems that operate the coaster, including control cabinets containing electrical components designed for operating motors and solenoids, operator panels, as well as the computer system that will operate and monitor the coaster.

The control system is defined as the electrical hardware and software that keeps the ride operating safely and smoothly. It communicates to the brakes, lift motor, queue gates, and restraint release systems how and when to operate. The control system also keeps track of each train's location and makes sure each vehicle is where it is supposed to be at all times. The design of an average roller coaster control system takes around 1,500 hours plus between two to six weeks to install and test.

Test and Adjust

Before a roller coaster can begin operating all the low zones must be blocked off. Areas where the track is low to the ground can be highly dangerous. Access to these zones must be blocked off with fences or walls and the exact requirements are defined in the safety standards. Despite the precaution of fences and warning signs, there have been numerous cases of trespassers or workers being injured or killed after being struck by a vehicle in a low zone.

The coaster enters the Test and Adjust Phase (T&A) after construction is complete and all electrical systems are operational. Operators anxiously wait as the coaster's empty cars race around the track for the first time. Will it clear all the hills as designed or stall out and become stuck? Imagine their nail biting as the engineers watch the cars crest the first hill. If the coaster makes it back to the station without stalling it has passed its first important test. But the control systems engineer's jobs are just beginning.

After the train makes it back to the station without incident several more test runs are made with sandbags or water dummies. Endurance tests are administered to detect any structural problems with the vehicle or track. The continuous running of the cars around the circuit is called cycling. Why is all this necessary? To keep you safe of course! Just like a car or your bike, a coaster must endure wear and tear. Tests ensure the coaster, as well as the track, stay strong year after year, rider after rider.

Strain gauges and accelerometers are utilized during this stage of testing the coaster. Strain gauges measure the actual loads applied to the undercarriages of the coaster's cars as they traverse the track. Strain gauges work by measuring a change in electrical resistance when the object is loaded. They permit direct measurement of the stresses in structural components.

Not only are the operators worried about the coaster and track breaking, but you too! Accelerometers measure bio-mechanical loads or the force your body is experiencing during your ride. An accelerometer is a damped mass on the end of a spring with a method for measuring the distance the mass moves in a particular direction. Operators use accelerometers to be absolutely sure that you'll come back to the platform in one piece having had the ride of your life.

Before the coaster can take on riders, all of the hardware, electrical, and computer logic must be tested. The block logic is tested by individually failing every proximity switch and photo-eye sensors to verify that the train will safely stop and in the correct locations. Brake speeds are tested and adjusted to ensure a smooth deceleration. Brakes are purposely failed to verify that the control system will receive the correct error and that it actually catches it. Good ride control programming can spread the wear over all the brakes, not just the first set. Every single system is tested extensively. The ride will be run for 100 cycles before taking on riders.

Government safety inspectors arrive to carry out a final inspection before giving approval to the park to operate the coaster. If all goes well the designer may get to ride on board so that he can assess the ride quality (and get his picture taken). In all, the ride will be ran hundreds if not thousands of times before it opens to the general public. During this period, the ride operators are trained by the engineers on how to safely operate the coaster.

What's the most common problem engineers run into when testing a prototype ride? It's always the one thing they didn't think would cause a problem. The second most common problem is the one issue they didn't think was even possible. Any problems need to be resolved as quickly as possible because opening day is fast approaching and the customer is planning on their large investment bringing in the crowds.

A roller coaster engineer does not always work 9am to 5pm like a regular desk job might. This is especially true when a new ride needs to be commissioned. Because many different departments need access to the ride to complete their tasks prior to opening, the schedule during crunch time may look like:

6am – noon: landscapers
Noon – 4pm: carpenters going through their punch list
4pm - 9pm: operations and crew training
9pm – 3am: engineering and controls

Operation

Operating a roller coaster safely and successfully requires a team of trained employees. The minimum number of people needed to run a ride varies per each ride and theme park. A typical coaster might have a restraint checker for the front and back of the train, a dispatcher standing at the control panel, and a grouper controlling the flow of guests into the station who also checks to ensure every rider is tall enough. Staff will rotate throughout their shift to stay fresh and to get breaks.

Once testing is completed and the crew has been trained the ride can be turned over to the operations personnel and guests are now free to ride. Roller coasters are designed for operational loads, non-operational loads, environmental loads, and operation in wind and non-operation in wind. Weather causes wood to swell or shrink. On a day-to-day basis, there are down-time reports generated that covers all of a park's rides and attractions. Any time a ride is non-operational, it gets logged on a report. The duration in which a ride is non-operational is called "down-time." There are typically three main reasons a ride or attraction is coded as down: Weather, Mechanical, or Operational.

Down-time due to weather is uncontrollable (unless the ride is inside). The roller coaster engineer must define recommended environmental restrictions for safe operation of the ride relating to environmental conditions such as, but not limited to, wind, rain, salt corrosion, and extreme heat or cold. Besides the obvious lightning strikes, weather conditions that may cause a ride to close are:

1. Wind velocity more than specified limits
2. Precipitation (rain, frost, snowfall, etc.)
3. Temperature below specified limits

Operational down-time is anything related to operator errors or guest issues (like bio clean up) and gets charged against the operations department. Finally, mechanical downtime is just that, anything resulting from a mechanical, electrical, or technical perspective causing a ride to go down and is charged against the maintenance department. Using this information, the park can monitor trends and issues pertaining to a certain ride or attraction and use it as a pro-active tool to keep the ride up and operating.

A roller coaster is no different than any other piece of machinery: it wears out over time. Compare this to a car, for instance. The more you drive the more your car wears, and to some extent, the more you pay to keep it up and running. The maintenance cost tends to be fixed whether your car is carrying one person or six people in the vehicle.

Imagine you're visiting an amusement park and the signature wooden roller coaster is running only one of its two trains. One train operation is frustrating to you as a guest because it means a longer wait time. However, for the park single train operation suggests the other train is sitting on the storage track. This is good for the operator because it means the wheels are not wearing, the bearings are not wearing, the brake fins are not wearing, the ratchet mechanisms in the restraints are not being used and wearing, etc. Every time a coaster takes a circuit, it wears a bit. Maybe the difference is only having to change the wheels out once a month instead of twice a month resulting in a significant cost saving. Multiply that by numerous coasters and it adds up fast. Eventually the maintenance costs become so great the park can't afford to operate it anymore. It may take five extra minutes for you to get on the coaster, but it could also result in the coaster being around for years to come. Keep that in mind the next time you are at an amusement park on a light attendance day and the park is not running their rides at full capacity.

Chapter 8: Safety

Safety is the number one priority in the amusement industry, and roller coasters are no exception. Coaster designers must complete risk assessments to identify possible modes of failure, their severity, and the likelihood of occurrence. Each failure mode is thoroughly examined to figure out how to prevent it or reduce its severity. Mitigation factors can be physical, like the anti-roll back devices in case the lift hill chain breaks, or they can be on the electronic/controls side, like using multiple sensors to detect things and making sure they all agree.

Every component that affects safety is either redundant or designed to fail safely. The computers used to control roller coasters are such that if there is any disagreement in the inputs, the ride shuts down. If one sensor fails, there is another one mounted in parallel to act as a backup. Friction brakes are designed to always be in the closed position. They require an input energy and the ok from the control system before opening to allow the vehicle to pass. Magnetic brakes don't require electricity to work in case of a power failure. Restraints are held in place by multiple hydraulic cylinders or ratchets. Even the wheels come in pairs.

One critical safety feature is the Operator Safety System (OSS). If you've ever rode a roller coaster you may have noticed all the ride attendants in the station holding down buttons before a train is dispatched. With all the guests in the station it may be possible for the ride dispatcher to lose track of all the ride attendants. The solution is to have designed safe points in the station platform. All attendants must be standing in the safe area before a train can be dispatched. By having buttons located in the safe zones the control system can account for all the staff members.

Unfortunately for amusement parks today, even when all the safety systems are working as designed, bad press can still occur.

News outlets try to overblow and sensationalize these non-stories for likes, comments, and shares on social media. In the headlines you'll see buzzwords like "malfunction", "roller coaster breaks", and "incident" when the safety system did exactly what it is designed to do: prevent a catastrophic accident from happening. Yes, sometimes riders must endure the inconvenience of being stuck in a stationary vehicle or climbing down the evacuation stairs, but these actions are rare and should only be viewed as a minor inconvenience as opposed to being seriously injured.

Amusement Industry Safety Standards

Amusement parks entertain millions of visitors every year, giving a countless number of safe rides. Visitors are much more likely to get hurt on their way to an amusement park then while riding a roller coaster. How is this incredible safety record achieved? There are no mandatory national safety standards for amusement rides, but the amusement ride industry has developed an extensive set of consensus safety and engineering standards through the ASTM International F-24 committee. (Fun Fact: ASTM used to stand for American Society of Testing and Materials but the American Society...International didn't make very much sense so now it's just plain ASTM International). These standards have been adopted or used as a model by state and local jurisdictions throughout the U.S. and around the world. In fact, thirty-five states reference ASTM industry standards in their regulatory laws for amusement rides.

The ASTM F-24 committee develops minimum standards of safety for design, manufacturing, operation, maintenance, quality assurance and inspection of amusement rides and devices, which includes waterslides, inflatables, go-karts, and zorbing devices. The committee consists of over 400 members, including ride manufacturers, amusement park and carnival operators, industry attorneys, safety consultants, ride inspectors, and regulatory officials. With the industry's full support and participation, the ASTM F-24 standards undergo frequent review and revision to keep up with the latest technologies. Biodynamic data is incorporated into the development process, which in turn produces amusement ride system guidelines that can safely accommodate the broadest segment of the population. ASTM sets the ride analysis standards.

The F-24 executive committee creates subcommittees, appoints members to leadership positions, and determines the scope and direction of standards development. Any member can apply to be involved in any subcommittee. The committee meets twice a year, in the spring and in the fall, to discuss revisions to standards. Attending meetings is a great way to network with engineers if you are interested in finding a job in the amusement ride industry. As a participant in the activities of the ASTM International Committee F-24 on Amusement Rides and Devices, you can help establish various standards on design and manufacture, testing, operation, maintenance, inspection, and quality assurance, all of which enhance the safety and integrity of the amusement industry.

There are three categories of ASTM Members: Producers, Users, and General Interest (Consumers). F24 consists of: Main Committee (F24), dealing with industry subjects, Subcommittees (F24.20) which look at specialized subject matters, and Task Groups (F2291) where individual documents are developed. There are six types of ASTM standards: Test Methods, Specifications, Practices, Guides, Classification, and Terminology.

The current list of all F24 subcommittees:

F24.10 Test Methods
F24.20 Specifications and Terminology
F24.24 Design and Manufacture
F24.30 Maintenance and Inspection
F24.40 Operations
F24.50 Training, Education and Certification
F24.60 Special Rides/Attractions
F24.70 Water Related Amusement Rides and Devices
F24.80 Harmonization
F24.90 Executive

Safety standards are highly scrutinized. Debates can range for hours just on proper terminology; whether a standard should use the word "shall" or "may." Is the standard a "guide" or a "practice"? Language like "properly use" in reference to safety equipment is vague and difficult to enforce. Thrill seekers raise their hands in the air on many rides and that is not a "proper use" of the amusement ride, even though a designer probably contemplated this in their design and made the clearance envelope large enough.

What does a standard look like? Standards are copyrighted and cannot be reproduced without permission but can be ordered from ASTM International's website (http://www.astm.org). As an example, wooden roller coaster operators or designers may have to follow guidelines like the following imaginary standard:

Section A12: Wooden Coaster Restraints

125.10.2 When restraints are provided by means of seat belts, the lap belt shall be installed so it engages the passengers at an angle with respect to a horizontal plane at 40 to 70 degrees.

125.10.2.1 When the passengers are properly seated and restrained the lap belt component of the restraint system shall not bridge over any component of the vehicle including but not limited to the frame, seat, and chassis.

125.10.4 Restraint belts shall be a minimum of 1 ¾" (44.45 mm) in width.

125.10.5 Restraint buckles shall have metal on metal latching mechanisms.

125.10.6 Inertia activated retractors, when used, shall be installed, and maintained according to the manufacturer's specifications.

Lock Out Tag Out Procedures

Lock Out Tag Out (LOTO) is a practice used to safeguard employees from unexpected startup of rides and attractions, or the release of hazardous energy during service or maintenance. Occupational Safety and Health Administration (OSHA) regulations require that only an authorized or qualified individual is to perform a Lock Out Tag Out operation. These rules require that all energy sources be turned off and either locked out or tagged out while service or maintenance work is being performed. The technician needs to de-energize the power source by setting a switch of some sort to the off position. Once the power source is de-energized, the technician will need to put a safety device on the switch, plug, valve, etc. The technician will need to place a pad lock and a tag (signed and dated by technician) on the safety device. Once the maintenance is performed, the technician will need to work in reverse order (removing lockout device and powering the equipment back on) and finishing the maintenance by testing the equipment.

The LOTO procedure should capture all valuable aspects that a maintenance person will need to know to perform the work. Ride Access Control (RAC) is a term used when authorized personnel performs certain tasks before powering down an attraction or equipment. This may include the following:

1. Equipment being de-energized
2. Location of equipment
3. Total Lockout Points
4. Description of equipment
5. Warnings
6. Required Equipment
7. Potential Hazards
8. Pre-Lockout procedures
9. Lockout Hazards
10. Lockout Procedure
11. Additional Information

12. Announce over the PA system that the ride will be shut down.
13. Lock gates.
14. Verify no personnel are in proximity of attraction.
15. Depress the E-stop.
16. RAC is also performed after attraction/equipment has been serviced, mentioned as "bringing the ride back to life."

Despite all the thought and planning that goes into LOTO procedures they don't work if they are not followed correctly. Every year in the amusement industry it seems there is always at least one news story involving the unfortunate maiming or death of a ride technician that could have been prevented had LOTO been properly implemented and followed. Indeed, there are roller coaster and amusement ride accidents caused by operator error. Despite the enormous effort and resources that both amusement parks and many other major corporations put into hiring, training, and monitoring their employees for safety and customer service, the system isn't perfect, and accidents do happen.

Inspections and Maintenance

Frequent visitors of amusement parks are probably aware that roller coasters are inspected daily. If you've ever been an early bird to the local amusement park you've probably seen the maintenance staff walking the rails looking for trouble. It's called preventative maintenance, which is coaster-speak for preventing accidents, keeping passengers safe all while they're shaking with fear. These troubleshooters are as important to the ride as are the thrills. They perform a multitude of tasks everyday meant to keep the coaster safely maintained and in top running condition. The goal of an inspection is to prevent accidents. If maintenance finds something worrying or downright dangerous the ride is going to be shut down, whether it's for a few minutes or an unfortunate few days, until they can fix whatever is ailing it.

Once a ride is sold to a customer, the ride designers no longer have control over its maintenance and upkeep. Although they create operation and service manuals, it can be difficult to verify whether or not the operator is following through on following the instructions. The design of the ride should try to account for this by not allowing a single failure to result in catastrophe.

Preventive maintenance is extremely important when it comes to thrill rides. But what does an inspection entail? There are a multitude of tasks that need to be performed on each area of the ride to keep it maintained and in great running condition.

First, before any inspections occur, the attraction must be locked out. This is tremendously important for the safety of the workers. There have been too many incidents in the amusement industry where a ride has been started with workers in dangerous areas leading to unfortunate and often deadly accidents.

A roller coaster's track is typically inspected every day before the park opens, at the request of management during the day, or due to an accident or near accident. Maintenance personnel will visually inspect high stress areas, such as the bottom of a big drop or the underside of the rails of airtime filled hills (read: times when you're screaming your head off). Before the coaster can be operated safely the track must be free of obstructions, such as tree branches, litter, cellphones, wigs, glasses, artificial limbs, etc. which might have fallen on the rails during the night or after a strong storm (all those items have in fact been found before). All the welds on handrails, stairs, and catwalks are meticulously examined for cracks. One of the common daily tasks is tightening loose bolts. On a behind the scenes tour, Dollywood's Pete Owns said that every one of Thunderhead's 250,000 bolts on the wood coaster sees a wrench at least once in the course of a year.

Most coasters require booster tires to move the trains from one block zone to another. These tires must be properly maintained and inspected for proper inflation and wear. This may include keeping the tire pressure at a specific level, such as fifty pounds per square inch (PSI).

There are several items to check on the lift hill including: chain dampeners, anti-rollbacks on the gearbox, sprocket and intermediate chain, chain trough, chain, and chain tension. In the station and queue areas the station air gates should be checked for proper alignment. All buttons and lights on the operator's panel should be working correctly. Transfer tracks should be locked in place. Brake shoes should be checked for excessive wear and proper alignment.

Next, it's time to examine the vehicles. The exact tasks vary from one coaster to another but typical daily inspections on the roller coaster cars include: inspecting the restraint system for proper operation, inspecting the condition of upholstery, condition of fiberglass, hitch yokes, safety cables, wheels for rotation, damaged urethane and proper oil levels, castle nuts securing wheels, chain dogs, anti-roll back dogs (ARBs), and shafts, axle center spindle for looseness, undercarriage of trains for cracks and missing safety wire for nuts. Excessive grease may need to be cleaned from each of the cars.

Before the ride can open to the public it must be cycled a specified number of times. The E-stop or emergency stop must be checked for the proper operation of the ride. This may include bringing the coaster's train to a complete stop on each of the ride's brake sections of track. After a visual inspection, the maintenance crew runs the roller coaster empty of passengers, watching and listening for any abnormalities. The crew may then take a spin themselves, this time paying attention to anything in the feel and the sound of the ride that seems out of sync. If a tire is flat, they'll be able to hear it. The roller coaster is open to the public only after a complete inspection of the cars, lap bars, and tracks that can take up to four hours or longer depending upon the size of the attraction.

The number of technicians inspecting each ride really varies on the size and complexity of the ride. A larger roller coaster will obviously take more time and personnel to inspect versus a kiddie ride. Different technicians have different job responsibilities, such as a mechanical, carpenter, electrical, and fiberglass. It takes roughly six man-hours to inspect the entire 3,230-foot track on Thunderhead, an average sized wood coaster. It takes three hours to inspect Merlin's Mayhem at Dutch Wonderland, a 1,300-foot suspended family coaster.

Ride components are checked for any signs of wear and often need replacing. There are two common terms when it comes to repairs. MTBF stands for the **mean time between failure**, meaning the average time between breakdowns of a ride or component. Ideally, this amount of time would be large as possible, like weeks or months or even years. If the ride does break down, then consider MTTR which stands for the **mean time to repair** or the average amount of time it takes to fix a ride. There's a proverb in the amusement industry that goes something like this: "If it's not broke, don't fix it. But if it is broken, you better be able to fix it in one night."

If a problem with the track is found it needs to be fixed right away. Dollywood spends up to $150,000 each year in track refurbishment and replacement for Thunderhead — and that's on top of standard wooden-coaster maintenance costs. Thunderhead received a new chain lift in 2019, the third new chain on the coaster since it opened in 2004. Chains typically last for 6-8 seasons.

Figure 65 – Notice the new wood on Thunderhead

These are just a selection of the tasks that need to be performed daily to keep an attraction properly maintained and running safely. There are other inspections that take place on a weekly, monthly, and annual basis. Or another method is basing maintenance on number of cycles: after 25,000 cycles do this...after 40,000 cycles inspect this, etc. A poor weld can contain microscopic cracks, invisible to the naked eye, so at least once per year amusement park operators X-ray their roller coaster's track or use magnetic scanners to check for metal stress or welds that need attention. Also, all vehicles are usually completely disassembled, inspected, and rebuilt at least once a year. Inspection schedules also vary between seasonal parks and parks open all year long. For example, seasonal parks will typically run all their trains all season long and then completely tear them apart for inspection during the off-season while other parks may buy one or two extra trains per coaster. That way, the coaster can operate at maximum efficiency while the extra trains are undergoing their yearly maintenance requirements.

Why are roller coasters removed?

Roller coasters are multi-million-dollar investments that amusement parks spend countless dollars and time in designing, building, and maintaining. But lately it seems like we've seen more roller coasters and classic thrill rides closed and dismantled than ever before. Why are roller coasters torn down? What goes into a decision to remove a costly investment, especially if it is a fan favorite?

Thousands of hours go into planning, designing, and engineering the perfect coaster, but it doesn't always go as planned or turn out as expected. This is especially true for coasters constructed in the days prior to advanced computer aided design. Design tolerances, the permissible limits in variation of dimensions and physical properties of manufactured parts, were larger than the tight engineering tolerances we can hold today. But try as they might, the engineers cannot account for every single variable.

Sometimes there are freak accidents, scenarios no one ever thought of. Or sometimes a ride is just plain boring and doesn't strike a chord with the public. Other times mother nature intervenes and usually not in a good way.

Regardless of the reason why, an amusement park ride may be closed but not dismantled right away. This status is called SBNO – Standing But Not Operating. Once a roller coaster is torn down, it is referred to as "defunct" meaning no longer existing. Let's look at some of the most common reasons why an amusement park might make the difficult decision to change the status of your favorite thrill ride from "operating" to "defunct."

Maintenance/End of Life

Just like any other material product, roller coasters have a shelf-life. Even if a ride is well-designed and well liked, sometimes the material just reaches the end of its endurance limits. When this happens, the track or components either have to be scrapped or replaced entirely. If almost every component needs to be replaced, the park may make the decision to demolish the existing ride, sell the material for scrap, and then build a brand-new coaster in its place. In Universal Orlando's case, this is what they did except the brand-new coaster has the exact same layout as the one that was scrapped. The Incredible Hulk at Island of Adventure had a complete replacement of all the steel including the track and support columns making it essentially a brand-new ride. Enthusiasts are already arguing over whether it should count as a new coaster credit or not.

Replacing a steel coaster structure is much more evident than a wood coaster. There's a saying in the industry: "you never stop building a wood coaster." Even though a classic wood coaster may have opened in 1950, how much of the structure is original? Sections of the ride may be replaced over time without you ever actually noticing it. Which begs the question, is it still the same coaster? It's a paradox known as "the ship of Theseus": the question of whether an object that has had all its components replaced remains fundamentally the same object.

Total Park Operating Cost/Budget

Contrary to what some amusement park goers think, theme parks cannot keep adding and adding roller coasters until they have fifty of them in the park – at least not without also increasing the attendance year after year thus increasing their revenue, no easy feat. Only so many coasters can operate at once without dramatically increasing ticket prices to offset the increased operational costs. A theme park can only have as many coasters as the operational budget allows. As Dick Kinzel states in his autobiography, amusement parks "break everything out by cost per rider." Rides with the highest cost per rider operating cost may be first on the chopping block.

Accident/Injury

Accidents often lead to ride closures because either A.) the coaster is not safe enough to operate and should be fixed or removed. Or B.) maybe it was a freak accident, and the ride is safe but the damage has been done: the ride and the operator's reputation has been hurt, maybe to the point where the public feels like the ride or even the entire theme park is unsafe and will no longer visit.

Real Estate/In the way of progress

"They don't have enough room to expand! How can they build another coaster? They're running out of space!" are cries often read on theme park forums about several parks. Usually, this complaint is baseless. Occasionally, though it is a legit problem. What's the easiest way to free up a lot of space in a theme park? Demolish an aging roller coaster. The land that the old coaster is sitting on becomes too valuable as a growing park runs out of room to expand.

Low or Declining Ridership/Unpopular

Another argument towards keeping or removing a roller coaster besides if it is adding/subtracting to the bottom line is how is it affecting the guest experience? Are guests coming to the park to specifically ride that ride? How disappointed will they be if they find out the coaster is broken down and closed for the day? Is the ride talked about in a positive or negative way? Is it hurting or enhancing the park's overall reputation? One way amusement parks will try to combat declining ridership is by modifying or adding to the attraction such as new theming, new train design, or adding VR headsets.

Natural Disasters

Occasionally, the decision to destroy a roller coaster may be taken completely out of the owner's hands, as in the case with natural disasters like fire, floods, earthquakes, and hurricanes. Hurricane Katrina resulted in the entire Six Flags New Orleans amusement park being permanently closed in August of 2005 (though it is being used to make movies, like Jurassic World).

Of course, the reason to remove a major coaster does not have to be mutually exclusive to one of the reasons but is more likely to be some combination of the above that all contribute to the owner pulling the plug. Or there could be another reason entirely. For example, the famous Cyclone Racer at the Pike in Long Beach, California was destroyed in 1968 because the City of Long Beach was attempting to improve its image and the coaster did not fit this new agenda. Regardless of the reason, us enthusiasts can only hope that the driving reason to remove a roller coaster is to build an even better one in its place.

Nick Weisenberger

Four Times a Roller Coaster Saved a Life

Despite often having a dangerous reputation due to the media sensationalizing the rare ride incidents, roller coasters are remarkably safe. In fact, some surprising health benefits of going to an amusement have been documented before. First, there's all that walking around a giant amusement park just to get to the roller coasters. When you go on the thrill ride, they're a good workout for your heart and lungs. Roller coasters are good for stress relief, fighting phobias, and clearing your sinuses. Some individuals have used them as motivation to lose weight, as Jared Ream did when he lost 190 pounds to be able to ride Orion at Kings Island. Riding roller coasters has even been found to be an unconventional way of clearing painful kidney stones.

But that's not all. While I always figured a day spent at a theme park is good for your health, I never realized it might also save your life. I've come across four instances where, had these individuals not ridden a roller coaster, they may not be around today to tell their remarkable, and eerily similar, stories.

I first heard about Emma Bassett's story in the book *Creating My Own Nemesis*, authored by famous theme park designer John Wardley. Emma Bassett had a huge brain tumor and was probably hours from death had she not ridden the Nemesis Inferno* roller coaster at Thorpe Park. The coaster ride redistributed fluid in her skull, relieving pressure on her brain that made it possible to live until she had two operations. John was stunned when he learned the news during a film shoot, a story he retells in his book.

*There are contradicting reports about which roller coaster it was Emma rode: Colossus or Nemesis Inferno.

Sally Dare was on a family vacation in Florida when she rode The Incredible Hulk coaster at Universal's Islands of Adventure theme park. Upon exiting the ride, she had headaches and blurred vision. The movements of the coaster had dislodged a tumor. Because of the early discovery, doctors were able to remove it successfully as it was only two cm in diameter. Without the coaster ride, it could have been another year or two before the tumor was discovered, and removing it might have been impossible

A visit to Thorpe Park in April 2016 saved Molly White's life, though she didn't know it at the time. When she got off Colossus, the ten inversion Intamin looping coaster, she had a headache. The enduring pain from the headache and vomiting fits made her seek out medical treatment which led to doctors discovering a brain tumor. Without riding the coaster and having the cancerous mass shifted enough to cause symptoms, who knows when it would have been discovered. It probably would have grown undetected until it was too late.

Michaeline Schmit was on a family vacation at Universal Studios in Orlando, Florida. She decided to ride a roller coaster and when she stepped off the ride, she had a migraine on the left side of her head. The pain never really went away, so once she returned home, she sought out medical treatment. After numerous doctor visits, she finally found out she had a brain tumor. Like the other women, if not for the roller coaster ride, it's possible the tumor could have gone undiscovered until it was too late to operate. Instead, Michaeline had a successful surgery and hasn't had any migraines since.

One of the lessons here is don't be alarmed if you have a headache immediately after riding a roller coaster or thrill ride. It doesn't automatically mean you have a brain tumor. But if you do have a migraine, dizziness, sight issues, or other unusual symptoms lingering several days after visiting a theme park, then you should seek out medical attention – it might save your life!

Every summer there's always at least one sensational story in the media making it seem as though roller coasters and amusement parks are unsafe. It's refreshing to hear stories from the other side. The next time you overhear someone in a queue talking about how "this ride killed someone!!" you can tell them about the times a roller coaster ride saved a life.

Chapter 9: Design Example

Engineers use physics and mathematics to design a safe and thrilling ride. In this example we're going to run a few calculations to compute:

1. The mass of the train.
2. The lift incline length.
3. The force required to pull the train up the lift.
4. The time required to reach the top of the lift.
5. The maximum speed of the train.
6. The radius of the bottom of the first drop needed to limit the g-forces on the passengers.

Please keep in mind that everything is generally simplified. Friction and other head losses have not been considered. For more understanding about the basic physics terms used such as velocity, acceleration, g forces, etc. please see the glossary in the appendix at the back of this book.

First, there are a few basic parameters that the customer has given us to start out with:

1. The height of the station track from the ground is ten feet or 3.048 meters.
2. The height of the first hill is 120 feet or 36.58 meters.
3. The angle of the lift is at 42 degrees.

We'll begin by calculating the force needed to pull the train up the lift hill incline and compute the power needed for the motor. To do so, we need to estimate the maximum mass of a fully loaded train to ensure that our lift hill motor can pull even the heaviest train of gravity loving coaster enthusiasts up the massive slope. To reach the hourly capacity target set by the client we will plan on having a six-car train. Each car holds two riders at 100 kg each, for a maximum mass of:

535 kg car + 2 x 100 kg riders = 735 kg.

735 kg per car x 6 fully loaded cars = 4,410 kg.

Therefore, the fully loaded coaster train will have a total mass of 4,500 kg (about 10,000 pounds). We selected the angle of the lift to be at 42 degrees. This means that the train is going to be pulled up vertically a distance of:

36.576 m - 3.048 m = 33.528 m.

The length of the incline will be:

33.528 m/Sin (42degrees) = 50.106 m.

We will finish calculating the force required to pull the train up the incline. There are two other assumptions we will make at this point: the velocity of the train as it exits the station and the velocity at the top of the lift hill. The velocity coming out of the station will be 10 mph or 4.4704 m/s. The speed at the top of the lift will be 18 mph or 8.04672 m/s. Energy is never destroyed; it is simply transferred from one body to another. Thus, we use an energy balance equation:

Kinetic Energy + Potential Energy + Work = Kinetic Energy + Potential Energy

Which is also written as: **KE1 + PE1 + W = KE2 + PE2**

Kinetic energy is a function of the velocity: $KE = (1/2)mv^2$.

Potential energy is a function of the height: $PE = mgh$

Work = Force × distance.

Substitute the KE, PE, and W equations into our energy balance equation and we get this resulting equation:

$$(1/2)mv^2 + mgh + Fd = (1/2)mv^2 + mgh$$

Now we can insert our values and solve:

$$.5(4,500)(4.4704^2) + 4,500(9.8)(3.048) + F(50.106) = .5(4,500)(8.046^2) + 4,500(9.8)(36.576)$$

F = 31,518.8 N of force or 7085.71 lbsf.

Now we know how much power the motor will need. One detail to note: we did not include the mass of the lift hill chain in our calculation. Plus, it's a good idea to overestimate and purchase a larger motor than needed.

How much time will it take for the train to reach the top of the lift hill? The acceleration of the train can be found using this equation:

$$(v_f)^2 = (v_o)^2 + 2ad$$

Inserting our values for final and initial velocity and distance:

$$(8.04672)^2 = (4.4704)^2 + 2a(50.106)$$

Now solve for acceleration and we get a= 0.45 m/s².

Next, use this equation to compute the time: **Vf = Vo + at**.

Input values and solve for t.

$$t = (8.05-4.47)/0.45 = 8.01 \text{ sec.}$$

That's pretty quick for a lift hill but it is only going up one hundred feet and at a 42-degree angle.

Now it's time to calculate the maximum velocity of the ride. Since the first drop is the longest, the velocity at the bottom will be the greatest. Energy relationships will be used to calculate the velocity:

$$KE1 + PE1 = KE2 + PE2$$

$$.5m(v1)^2 + mgh = .5m(v2)^2 + mgh$$

Solve for v2 and we get 27.42 m/s or 61.34mph!

Finally, we want to figure out what the radius of the curve of the bottom of the first drop should be to keep the g forces felt by the riders to be 2.5 g's or less.

$$g's\ felt = g's + 1$$

$$2.5 = g+1,\ g = 1.5$$

$$g's = Acentripetal/9.8$$

$$Acentripetal = 1.5 \times 9.8 = 14.7$$

$$Radius = v^2/Acentripetal = 27.42^2/14.7 = 51.14m\ or\ 167.78\ feet.$$

Congratulations, you've just taken the first steps to becoming a roller coaster engineer! Do real roller coaster designers do all the math by hand like this? Not necessarily, but to be an engineer you need to understand the theory behind it. Software is simply a tool to make the job easier.

Chapter 10: Career Advice

So, you want to be a roller coaster designer? Well, who doesn't? The field is highly competitive, and most roller coaster engineering firms are very small. One company, for example, was producing about one coaster every year and only employed a total of four engineers! When working for a small company (and by small, I mean less than fifty employees), no matter the field, the responsibilities of the individual employees are amplified much more so than at a large corporation. Each team member's time and skills are utilized to the maximum. No one engineer does purely design work. Long hours, cross country travel, and days away from home come with the job. It's not for everyone.

If you ask a roller coaster engineer "What is a typical workday like?" the answer will likely be there is no typical workday. An engineer's role can vary significantly from one day to the next. Monday could be spent on creating structural drawings, Tuesday on recreating an existing ride from survey data for a repair job, Wednesday on putting together a new layout concept for a potential customer, Thursday on writing up and signing purchase orders for components, and Friday on stress analysis of the structure.

The technical approach to becoming professionally involved in roller coaster design is to obtain an engineering degree then get a job with a ride manufacturer or a specialty consulting firm. Mechanical and civil engineering are the two dominant majors in the industry. Structural and electrical are common too though you are certainly not limited to those. As an engineer it is very rare that you will work exclusively within your discipline. Mechanical engineers today are more like electro-mechanical engineers who also do civil engineering work. Pay is competitive to similar engineering positions in other industries.

Pursuing a career in roller coaster engineer does not mean you have to work directly for one of the well-known coaster design firms either. Coaster engineering companies may outsource work to smaller suppliers. It's not all track and car design. There are other components that need to be designed and manufactured, like the station gates that automatically open and close each time a train pulls into the station. All the control systems and electrical work is usually outsourced to other companies that specialize in such business. You could repair old or existing rides or get a job in sales trying to sell the next one.

What school or college is the best for roller coaster engineering? It doesn't matter. Really. You could attend the best engineering school in the world but if you don't apply yourself and put in the time and the effort then it will amount to nothing. Attitude and aptitude carry greater weight than the school you attend. Some employers may favor candidates from one college over another, but the bottom line is they are all looking for someone who truly understands the material and has a passion for the industry. The few companies that do have internships usually expect applying students to have at least a 3.0 GPA. But don't fret over it too much, engineering companies are often more interested in a person's skills and experience rather than a high GPA or a prestigious school. It's more important to show how you can provide value to a potential employer. Ability to communicate, go above and beyond, and a willingness to open up are key.

It would, however, be a good idea to attend a school that has an established theme park student group. But if your school doesn't already have one then you can be the one to take the initiative and start it, which will be a great talking point in a job interview.

Engineering companies today like to see new employees with at least six months of actual work experience at an amusement park under their belt, especially working hands on with rides. Of course, a current college student or recent graduate may not have a lot of actual work experience but there are other ways to showcase your passion and skills. A proven knowledge of the industry or a home-made project shows an interest in the subject matter. Any work related to the field or market helps even it is not in a technical area.

To have the best chance of getting a full-time job in the industry, you need to set yourself up for success. The best way to do that is by getting involved and making connections. Nowadays, there are many online or in person conferences and design competitions, sponsored by universities or companies, that allow students to show off their skills and get a taste of the amusement industry. A few worth looking into:

SKYnext, hosted by Skyline Attractions, is an event for young professionals looking to get into the amusement industry. SKYnext offers exclusive presentations, behind-the-scenes tours, and networking opportunities with other passionate individuals.

Ohio State University's Theme Park Engineering Group has hosted several conferences designed for students looking to pursue careers in the themed entertainment and amusement park industry. **Students in Themed Entertainment**, or **S.I.T.E.**, will give you an inside look at how Themed Entertainment and TEA-affiliated clubs throughout the nation contribute to the world of Themed Entertainment. You will get the opportunity to see the projects and skills university students from all over the world have been working on in preparation for joining the industry full-time. And no, you do not need to be an engineering major to be involved in the theme park engineering group.

The Toronto Metropolitan University Thrill Design Invitational Competition (formerly Ryerson Invitational Thrill Design Competition, RITDC), presented by Universal Creative, is a valuable experience and exposure for students representing dozens international themed entertainment clubs. This competition was across four days hosted at the Universal Orlando Resort. Teams were presented challenges on-site and were tasked with creating a solution and presenting their ideas within 24-48 hours to a panel of judges of senior leadership in Universal Creative and some of their partners. Disney's Imaginations Design Competition is similar.

Themed entertainment projects are so rewarding because of the people you get to work with and how their individual passions contribute to the overall success of the project, which is missing in an individual/personal project; however, one of the cons of RITDC was that content created for the competition is considered owned by Universal and confidential. So, in job interviews you can only really describe some of the problems solved and can't showcase it.

Another way to showcase your skills is by sharing a personal project. Personal projects can be hard because they are self-starting, meaning you must find the discipline to complete the work since there are no actual deadlines or someone managing your results. If possible, find someone or a group to share updates of your work with, or find ways to motivate certain milestones.

Roller coaster engineering jobs are few and far between, so if you're going to get an engineering degree so you can design coasters, you better really enjoy engineering first. If you don't get a coaster design job right away, you'll want to be at least somewhat happy as a "non-coaster" engineer. If you don't look forward to the class work, then the profession may not be for you. It's also worth noting even roller coaster engineering firms may take on additional engineering work to help balance their books, such as foundation work, bridge design, ski lifts, or other transportation jobs.

Believe it or not, it is possible to go through several years of college and get a degree, even with good grades, and still not really understand the subject that is featured prominently in the middle of the diploma. Truthfully, you may never have to think about eighty percent of what you learn in college. But the unexpected twenty percent of what you learned will suddenly become a large and important part of your work. If you are one who does not like what you are studying enough to look forward to your class work, you ought to find another field of study. An interviewer will know within a few questions if you have any real interest and aptitude for the subjects you claim to have studied. To design thrill rides you must be an engineer first and a roller coaster designer second. You must love engineering, and if you happen to find yourself engineering roller coasters one day then consider yourself lucky!

Of course, engineering is not the only avenue one can take to be involved in thrill ride development. Theme park attractions are assembled by a large team of people with skills ranging from design and development, to operations, set design, planning, construction management, marketing, and finance. For example, the artistic approach could be where you get experience in theater, television, or film to eventually be hired in the art department of a theme park. Jobs include concept art, marketing materials, set design, theme design, or model maker.

Is it very hard to get into this profession? Yes. Is it impossible? No. Don't fall into the "I wish I knew what I know now when I was younger" category. There are a number of strategies you can implement to improve your chances of finding a role within the amusement park industry (in no particular order). Don't rely on luck; make your own luck.

If you're serious about getting a job in the amusement park industry then you should attend the International Association of Amusement Parks and Aquariums (IAAPA) yearly conference, usually held in Orlando, Florida (or attend the European or Asian equivalent if you're not based in the United States). There are thousands of attendees at this event from all corners of the globe. Park representatives are on hand to try and sell or purchase new amusement park rides or attractions. This is a great place to meet people and learn more about the industry. It can't be overstated how important making connections within the industry is.

You should also seriously consider joining the ASTM International F-24 Committee on Amusement Rides and Devices. ASTM holds a bi-yearly conference to discuss the creation of safety standards and practices. The ASTM conferences have a more relaxed atmosphere compared to IAPPA. There's no pressure on the employees to sell any rides and everyone is generally totally relaxed, open and friendly. Not to mention nearly every attendee seems to have vice president, director, manager, or president attached to their job title. A great number of participants will be more than willing to take a few minutes of their time to talk to you about the industry and possible opportunities for you. If you're looking into job openings in ride design or engineering, you need to join the ASTM and participate in these meetings.

Now, don't go to ASTM just to pass out your resume. The point is to learn and help improve the industry, but it is also a great way to earn respect and get noticed. Talk to the other attendees and if they ask for your resume be prepared to give it to them or be sure to get their business card to email them later.

Some coaster manufacturers even make efforts to interview potential interns, share industry knowledge, facilitate networking between fellow students interested in the amusement industry, and provide access to industry professionals through guest speaking. Clubs like the American Coaster Enthusiasts (ACE) occasionally sponsor tours at coaster design firms. These are a great opportunity for you to introduce yourself and get on their radar.

Often, a professional organization that requires a yearly membership fee will have a student discount or separate student club. For example, the Themed Entertainment Association (TEA) has a NextGen program for students or recent graduates from university, community college, or other institution. Your membership provides you the opportunity to develop new professional contacts to help grow your career.

As far as job interviews and resumes, think about what makes you different from everyone else trying to get into the industry. How does that make you a better candidate? Make yourself standout. Employers seeking a new employee want to see ambition and a positive work ethic that will translate into productivity. This can be presented in grades, work history, achievements, projects, and in the interview itself. You need to show you're a team player who can effectively communicate. How will you add value to their team?

Use your resources, especially those online resources such as LinkedIn.com. It's a great way to get introduced to professionals in the industry. Join groups and participate in online discussions to gain respect and showcase your knowledge and skills. Use social media to your advantage. Make a portfolio of your work. Keep in contact with industry professionals you meet and never, ever, EVER burn any bridges. Like ever. It's a small world.

Also be aware of confidentiality. Coaster designers cannot release any information related to a project they are working on until that information is released to the public by the customer. So, it's probably not a good idea to ask a coaster designer what they are working on or trade secrets because they probably can't tell you.

A word of warning: engineers must have thick skin. There are times when you will put in countless hours designing a part or a scenic element when at the last minute the decision is made to cut the part from the project (most likely due to budget) and your first thought will be "All my time and effort the past couple of weeks has been completely wasted." But you can't get mad or upset. Understand this is a common occurrence in the word of engineering and simply move on to the next task. There is too much to do to spend time worrying about things beyond your control.

As the famed theme park designer John Wardley once stated, "Often an intense enthusiasm for a subject gives one a false perspective of the true nature of the industry behind it." Even if you end up designing coasters, you'll more than likely still be working in an office environment, the main difference is you'll be doing paperwork for roller coasters instead of paperwork for car parts, printers, or air compressors. It's not all about riding roller coasters or playing computer games all day.

This is not meant to discourage you from pursuing a career in roller coaster design; it's simply a realistic viewpoint to help you manage your expectations. Roller coaster engineers will tell you the work can be quite challenging but, in the end, when guests are stepping off the ride laughing, it is very rewarding. The job sure does have its ups and downs.

Another route to try if you just want to be involved in the amusement industry in some way is ride inspector. You'll get up close and personal with all rides, from major amusement parks to small county fairs to inflatables at the local carnival. Here in my home state of Ohio, the Ohio Department of Agriculture's Amusement Ride Safety Division inspects them all. The duties for an Amusement Ride and Game Inspector position might include:

- Independently inspect amusement devices of domestic or foreign manufacturers for safety and compliance with manufacturer's design and standards (e.g., checks adequacy of structure and moving parts by climbing on, around, under, and over structure, determines adequacy of passenger carrying devices for safety, appearance, and comfort, reads manufacturer's specifications and blueprints to ensure all characteristics of ride follow the manufacturer's specifications)
- Determines whether amusement rides are to be licensed
- Cites violations and takes necessary corrective action which includes temporarily revoking license of ride owner and conducts investigations of accidents occurring on rides
- Conducts mid-season operational inspections
- Conducts inspections of fairgrounds, games and novelties to ensure games are licensed in accordance with applicable state laws and rules and to protect public from use of illegal devices, unscrupulous games and concession owners at county, independent and state fairs (e.g., determines whether games are operated as game of skill or chance).
- Attends seminars, training sessions and engages in individual studies to upgrade ride technology
- Prepares required reports
- Meets with members of fair boards, owners and concessionaries to explain laws and rules pertaining to concession operations
- Conducts public awareness ride safety sessions with school age children, owners, operators and city officials.

Scores of professionals have stumbled into the amusement industry by accident. You never know when your opportunity is going to come but when it does be prepared and be ready! It takes work and persistence, and perhaps a bit of good timing and luck, to break into this industry. Hopefully this advice helps you achieve your goals of becoming an active member of the amusement park industry. Good luck and never give up!

Career Advice Summary

Final thoughts for reaching your goal of becoming a theme park engineer or roller coaster designer:

- ❖ Attend the International Association of Amusement Parks and Aquariums (IAAPA) yearly conference.
- ❖ Join ASTM International F-24 Committee on Amusement Rides and Devices and attend one of the bi-yearly meetings.
- ❖ Join a school group such as the Ohio State University's Theme Park Engineering Group. If your university doesn't have one, consider starting it yourself. Reach out to leaders at colleges that do have one if you need advice.
- ❖ Work at an amusement park in any capacity, even if it's just flipping burgers.
- ❖ Compete in a themed entertainment ride design competition
- ❖ Work on and showcase a project (3D printing, art concepts and sketches, NoLimits coaster designs, etc.)
- ❖ Network, Network, Network! Online and in-person.
- ❖ When in school, don't be afraid to ask for help when you need it.
- ❖ Be a well-rounded person, the more skills you can bring to the table, even outside of engineering, the better.
- ❖ Study and work long and hard.
- ❖ Communicate effectively.
- ❖ Get your name out and get on your dream company's radar.
- ❖ Maintain the right attitude.
- ❖ Keep knocking on doors.
- ❖ Always be the best you can be. Ask yourself, "Is this the best I can do?"

Coasters 101: An Engineer's Guide to Roller Coaster Design

Chapter 11: Enjoy the Ride

There you sit, barred, and buckled into a scream machine climbing to 208 feet, the amusement park far below, the fear palpable as your car steadily approaches the summit. In the next instance you're over the apex, plummeting to the ground at breakneck speed. The air rushes past your face stifling your screams, you feel your body being compressed into the seat, the infrastructure whips past you in a blur.

One minute you're falling, the next you're rising towards the heavens again. The car ascends a tall hill, quickly losing speed, and you have a quick moment to catch your breath. At that awesome instance your eyes drift warily over the giant's superbly crafted metal structure, and you marvel at the skill required to design and assemble such perfection. The thought is fleeting as the floor suddenly drops out beneath you as the monster hurtles you back towards the Earth. And then it's all over. You're safely back on solid ground, out of breath, wondering what just transpired. You unhook the safety belt, climb out of the car, and exit the station. Then you get right back in line, ready to do it all over again! The engineers have done their job well.

When a roller coaster designer beholds their completed ride, they often can't help wondering about what might have been. To the average rider, there is only one coaster, the one that he or she is waiting anxiously to ride. But just like the completed coaster, the design process had its own ups and downs. Its evolution included multiple configurations of the ride's path. The engineers may even get nervous taking a turn on it because they knew a bent or chain lift engine housing once collided with the ride path in the 3D model before being modified. But as loopy as the creativity gets or the steeper the drop, or the faster the turns and the louder the screams, one thing they aren't, overall, is unsafe.

Roller coasters are sophisticated systems that are a thousand technical miles beyond the Russian ice slides they evolved from. The amusement industry has entered an era of high-tech engineering. Higher speeds, higher loads (both structural and biomechanical), faster acceleration, crazy inversions, cantilevered seats, spinning cars – the designs are becoming increasingly more complex. Whether it is a deliberately shaky, wooden ride or a corkscrewing demon with legs dangling in mid-air, every midway monster is a complex three-dimensional puzzle to be solved by the engineers.

Roller coasters are a perfect mix of engineering and art, and designers are constantly pushing the limits. The next time you ride a roller coaster you should really take a minute to stop and appreciate that every single component was modeled, drafted, and constructed by someone. It's easy to look at a roller coaster and be impressed with the sheer size of the structure and the insane number of parts. It's much harder to comprehend that someone had to put each one of those pieces together, both conceptually and physically.

Roller coasters exist to provide us with a break from everyday life. They create an exhilarating - and often addicting - distraction from the experiences your senses are used to. The feeling of being out of control without any real danger is not easy to come by. Humans, by nature, are programmed to go looking for danger. Skydiving, mountaineering, and racing are just a few of the many sports man indulges in just for the thrill of it. Not everyone is such a daredevil, however, and with the use of appropriate technologies, people are now enjoying these dangers without the dangerous part. Roller coasters are a perfect example of such ingenious use of technology. Pick your amusement park, anywhere on the planet, and where are the longest, most aggravating lines? The roller coasters, of course! They are the main attractions on the midways these days, with millions strapping themselves into these scream machines because they love being scared in a safe environment. The success of a ride can easily be measured in smiles!

The next time you climb aboard, think about all of that and appreciate the thought, time, energy, and resources that go into designing these amazing thrill rides. Then put your hands in the air, scream at the top of your lungs, let loose, and enjoy the ride!

-Nick Weisenberger

Figure 66 – Racer at Kings Island

Nick Weisenberger

Did You Like *Coasters 101: An Engineer's Guide to Roller Coaster Design*?

Before you go, I'd like to say "thank you" for purchasing my book. I know you could have picked from dozens of other books, but you took a chance on mine. So, a big thanks for ordering this book and reading all the way to the end.

Now I'd like to ask for a *small* favor. Could you please take a minute or two and leave a review for this book on Amazon.com? Your comments are valuable because they will guide future editions of this book and I'm always striving to improve my work.

Figure 57 – Zamperla spinner Tidal Twist at Columbus Zoo and Aquarium

About the Author

Nick Weisenberger has been interested in theme parks as long as he can remember. His childhood love of trains evolved into a fascination of roller coasters, thrill rides, or anything else that ran on rails. Perhaps it was easy to spark interest in a child raised in central Ohio, just hours away from some of the best roller coasters in the world. Ironically, he remembers loving roller coasters long before he had the courage to ride them.

As a kid, Nick would build huge roller coaster tracks all around the house using plastic K'nex building sets, attempting to build the longest and most exciting coaster that gravity and friction would allow. He not only rode his sled down the hill in his backyard during the winter, but also constructed a giant, banked turn that would allow for his ride to be more like a bobsled. Nick rode cardboard boxes down the stairs. He first began playing coaster-design computer games in 1993 on a DOS system. Yes – he admits he was exceptionally nerdy and coaster crazy. And now he is an engi-nerd!

Nick has been employed by an engineering services company where he participated in projects ranging from the automotive and aerospace industries to amusement parks and roller coasters. His work included 3D modeling and kinematics simulations, stress analysis, 2D to 3D data conversion, updating lock-out tag-out procedures, maintenance planning efficiency, and more. He even designed several 3D CAD models and simulations for a never-been-built-before roller coaster concept.

Nick is currently co-manager of Coaster101.com as well as a member of the ASTM International F-24 committee on Amusement Rides and Devices. In August 2009, he participated in the Coasting for Kids Ride-a-thon where he endured a ten-hour marathon ride (that's 105 laps) on Gemini at Cedar Point and helped raise over $10,000 for Give Kids the World charity. His favorite roller coaster to ride is The Voyage at Holiday World in Santa Claus, Indiana for its extreme airtime and out-of-control feeling. You can occasionally catch Nick on episodes of the Coaster101 podcast. An avid traveler, look for Nick on the midways of your local amusement park.

As of October 2022, Nick has ridden 265 coasters. He knows exactly how many because he uses a spreadsheet to keep track. If you'd like to track your own coaster count, you can download a copy of his spreadsheet template here: https://gum.co/coastercount

Figure 68 – The author at the opening of Kennywood's Sky Rocket

Special Thanks - I'd like to personally give a special thanks to everyone who either helped directly or inspired me to write this book: Adam House, Adam Sandy, Andrew Stilwell, Bob Gurr, Brendan Walker, Camiel Bilsen, Corey Rasmussen, Dal Freeman, Don Helbig, Dr. Kathryn Woodcock, Eric Wooley, James St. Onge, Jeff Pike, Joe Cornwell, Joe Relich, John Hogg, John Stevenson, John Wardley, Josh Adams, Katherine Johnson, Kyle Lindner, Larry Treece, Mark Stepanian, Matt Schmotzer, Mike Martin, Nathan Unterman, Neil Wilson, Paul Gregg, Pete Owens, Phil "The Ride Guru" Bloom, Scott Rutherford, Shane Joseph, Steve Boney, Tyler Mullins.

The 50 Most *Terrifying* Roller Coasters Ever Built

Mega roller coasters of today reach heights of over 400 feet and speeds more than 100 miles per hour. Roller coasters towering taller than a certain height are terrifying for many individuals, but it would be boring to simply make a list of the world's tallest coasters. As a result, most of the bone-chilling machines in this list do not use sheer height to terrify, but instead prey on our fears and emotions in other, more creative ways. One element alone may not make a ride terrifying but the sum of all its parts does.

What factors make a roller coaster terrifying? Height, speed, inversions, backwards segments, unique track elements, darkness, and unexpected surprises all contribute to making your head spin and your knees tremble. Where are the most terrifying roller coasters found? Who designs them? Which park builds the craziest rides? Find out by reading ***The 50 Most Terrifying Roller Coasters Ever Built!*** Look for it on Amazon.com today.

"This is a fantastic book that gives great insight and ideas of where to travel for harrowing and fun-filled roller coaster experiences - it gave me an adrenaline rush just reading about the possibilities."
–Amazon reviewer

"My roller coaster loving son thoroughly enjoyed this book. Easily broken down by ride with fun and pertinent facts about each coaster." – Amazon reviewer

Works by Nick Weisenberger

25 Extreme Drop Tower Rides

50 *Ground-breaking* Roller Coasters

50 *Legendary* Roller Coasters That No Longer Exist

A Brick-by-Brick Guide to Legoland New York

Coaster Phobia: How to Get Over Your Fear of Roller Coasters

Coasters 101: An Engineer's Guide to Roller Coaster Design

Steam Trains and Monorails

The 50 Biggest Ferris Wheels Ever Built

The 50 Most *Terrifying* Roller Coasters Ever Built

The 50 Most *Unique* Roller Coasters Ever Built

Things to Do in the Smokies with Kids

Appendix I: Acronyms

The following is a list of acronyms found within this text and includes common terms used throughout the amusement industry (in alphabetical order).

AC: Alternating Current
ACE: American Coaster Enthusiasts
ARB: Anti-Roll Back
ASD: Allowable Stress Design
ASTM: American Society of Standards and Materials
CAD: Computer Aided Design
CARES: Council for Amusement Recreational Equipment Safety
CATIA: Computer Aided Three-Dimensional Interactive Application
COG: Center of Gravity
CPM: Critical Path Method
DC: Direct Current
DOT: Direction of Travel
ESPE: Electro-sensitive Protective Equipment
FEA: Finite Element Analysis
FEC: Family Entertainment Center
FMEA: Failure Mode and Effects Analysis
FTA: Fault Tree Analysis
FVD: Force Vector Design
GDT: Geometrical Dimension and Tolerance
HAZ: Heat Affected Zone
HMI: Human Machine Interface
IAAPA: International Association of Amusement Parks and Attractions
ISO: International Organization for Standardization
LED: Light Emitting Diode
LIM: Linear Induction Motor
LOTO: Lock Out Tag Out

LPL: Large Project Leader
LRFD: Load and Resistance Factor Design
LSM: Linear Synchronous Motor
MBD: Model Based Definition
MTBF: Mean Time Between Failures
MTTR: Mean Time To Repair
NDA: Non-disclosure Agreement
NDT: Non-Destructive Test
OEM: Original Equipment Manufacturer
OSHA: Occupational Safety and Health Administration
OSS: Operator Safety System
PERT: Project Evaluation and Review Technique
PES: Programmable Electronic Systems
PIC: Person-In-Charge
PL: Project Leader
PLC: Programmable Logic Controller
POV: Point of View
RA: Ride Analysis
RAC: Ride Access Control
ROI: Return on Investment
RPM: Rotations Per Minute
RSS: Ride Show Supervisor
SARC: Standard Amusement Ride Characterization
SAT: Site Acceptance Test
SBNO: Standing But Not Operating
SLC: Suspended Looping Coaster
SRCS: Safety Related Control Systems
T&A: Test and Adjust
THRC: Theoretical Hourly Ride Capacity
USPTO: United States Patent and Trademark Office
VFD: Variable Frequency Drive

Appendix II: Glossary

The following is a list of common physics and engineering terms used throughout this text, with definitions:

4th Dimension: A type of roller coaster with controlled rotatable seats cantilevered on each side of the track.

Acceleration: The rate of change of velocity with respect to time. If an object is speeding up, slowing down, or changing direction, it is accelerating (or decelerating). Acceleration describes the rate of change of both the magnitude and the direction of velocity. Acceleration = Force divided by mass **(a=F/m)**. Impact acceleration are those with durations less than 200 ms. Sustained acceleration are those greater than or equal to 200 ms.

Accelerometer: An electromechanical device that will measure acceleration forces.

Airtime: Roller coasters can thrust negative g's on riders causing them to momentarily lift off their seats and become "weightless." As the vehicle flies over the top of a hill the load on the passenger becomes less than Earth's gravity and, in the extreme, could throw an unrestrained passenger out of the car. Scream machines with oodles of so-called "airtime" moments or "butterflies in your stomach" thrills rank among the world's best. Negative g-forces cannot be too great because when under high negative g forces blood rushes to the head and can cause "red out."

Amp: The SI unit for measuring an electric current is the ampere, which is the flow of electric charge.

Anthropometry: The study of the measurements of the human body.

Balking: When a guest decides not to wait in line for an attraction due to the wait time.

Block: A block is a section of a roller coaster's track with a controllable start and stop point. Only one train may occupy a block at a time.

Bobsled: Cars travel freely down a U-shaped track (no rails) like a bobsled except on wheels.

Bunny Hops: A series of small hills engineered to give repeated doses of airtime.

Capacitor: A device used to store an electric charge, consisting of one or more pairs of conductors separated by an insulator.

Catwalk: A narrow walking ledge with handrails, often found along brake runs or lift hills to help with maintenance and evacuations.

Centrifugal Force: The force that pushes something moving in a circle towards the outside edge.

Centripetal Force: The component of force acting on a body in curvilinear motion that is directed toward the center of curvature.

Charge: In general, charge Q is determined by steady current I flowing for a time t as $Q = It$.

Cobra roll: A half-loop followed by half a corkscrew, then another half corkscrew into another half-loop. The trains are inverted twice and exit the element the opposite direction in which they entered.

Corkscrew: A loop where the entrance and exit points have been stretched apart.

Cycle: When the train completes one circuit around the course. When trains are run continuously this is called cycling.

Cycloid: The locus of points generated by a fixed point of a circle as the circle rolls along a straight line.

Derailment: When a roller coaster train or vehicle leaves the tracks.

Dive loop: A roller coaster element where the track twists upward and to the side, like half a corkscrew, before diving towards the ground in a half-loop. Basically, the opposite of an Immelman inversion.

Dueling Coaster: Two separate tracks but mostly not parallel. Usually contain several head-on, near miss collision sensations.

Drive tire: Drive tires can accurately control the position of a train.

Dwell time: The time between when one train or vehicle exits the station until the next one is ready to load passengers. The longer the dwell time the less efficient the ride operation can be.

Dynamics: The study of the causes of, and changes in motion.

Ergonomics: The study of people's interaction with the workplace.

Element: A segment of coaster track that curve in a recognizable shape. Elements are often given names such as corkscrew, loop, cobra roll, dive loop, etc.

Energy: Energy is the ability of one system to do work on another system. Energy cannot be created or destroyed.

E-Stop: Shutdown sequence other than a normal stop and may be more dynamic. Also known as emergency stop.

Floorless coaster: The vehicle sits above the track but contains no floor between the rider's feet and the rails, allowing their legs to dangle freely.

Flywheel: A flywheel works by accelerating a rotor to a very high speed and maintaining the energy in the system as rotational energy.

Force: An influence on an object which causes a change in velocity, direction, or shape. Force equals mass times acceleration (**F=ma**).

Force Vector Design: A method of roller coaster design using mathematical formulas. This system inputs a series of g-force formulas and time intervals and generates a perfect ride path based on the desired effects.

Friction: A force that resists relative motion (sliding or rolling) of an object.

Full circuit: An uninterrupted closed loop path or track.

Giga coaster: Any roller coaster with at least one element between 300 and 399 feet tall.

Gravity: The force that tends to draw objects toward the center of the Earth.

Gauge: Distance between the center of each rail on a coaster's track. Also called the pitch.

Hill: A concave shaped element on a roller coaster's track.

Hyper coaster: Any roller coaster with at least one element between 200 and 299 feet tall. Magnum XL-200 at Cedar Point was the first.

Imagineer: A person who works for Walt Disney Imagineering. This term is a combination of engineer and imagination.

Immelman: Named after the aircraft maneuver pioneered by Max Immelman, the inversion begins with a vertical loop but at the apex of the inversion turns into a corkscrew exiting at the side instead of completing the loop. The opposite of a dive loop element.

Inertia: The tendency to resist a change in motion. If an object is moving, it won't slow down or change direction unless acted upon by an outside force, such as friction.

"An object at rest stays at rest and an object in motion stays in motion with the same speed and the in the same direction unless acted upon by an unbalanced force."

To explain this, let's think about an environment free of gravity, like space. If you throw a ball in space, it will theoretically fly forever in the exact direction and the exact same speed with which you threw in. However, if it met an unbalanced force such as a meteor, it would change its direction.

Inversion: An element on a roller coaster track which turns riders 180 degrees upside down and then rights them again, such as a loop, corkscrew, or barrel roll (among others). The actual degree of banking to be considered inverted is debatable, with some arguing 135 degrees is closer to upside down than not and should be considered an inversion.

Inverted coaster: Vehicle is fixed below the rails with rider's feet hanging freely and can invert upside down. The terms "inverted" and "suspended" are often used interchangeably by parks.

Jerk: The rate of change of acceleration, expressed as m/s^3

Kinetic Energy: The energy an object possesses due to motion. Kinetic energy is one-half times the mass multiplied by velocity squared (KE=0.5*m*v*v).

Laydown/Flying coaster: Riders are parallel to the rails, either on their back or stomach.

LIM: Linear Induction Motors are a form of electro-magnetic propulsion. The stator and rotor are laid out in a line which produces a linear force (as opposed to a torque or rotation). No moving parts.

Manufacturer: The party producing the amusement ride or device. Can include the designer/engineer.

Mass: The amount of matter within an object. The terms mass and weight are often used interchangeably. However, weight = Mass * Gravity thus weight can be zero when no gravitational forces are acting upon it while mass can never equal zero.

Mobius: A racing or dueling roller coaster with one continuous track instead of two separate ones.

Momentum: Momentum is the product of the mass and velocity of an object (**P=mv**). The larger the mass or velocity an object, the more momentum it will have.

Motorbike coaster: Riders straddle the seats as if riding a motorcycle, jet ski, or horse. Also called straddle coaster.

Neutral slope: The amount of downgrade on a roller coaster track that will maintain an average speed of a car in motion without any change in velocity.

Normal Force: The force component perpendicular to the surface the object rests on. In most cases, the normal force (N) equals mass times gravity (**N=mg)** or if the object rests on an incline

$N = mg\cos(\theta)$ where theta is the angle to the inclined surface measured from horizontal.

Off-axis airtime hill: An airtime hill where the peak is banked at an angle rather than parallel to the ground, giving negative and positive lateral gs at the same time.

Patron: The customer who is experiencing the attraction. Also called rider, guest, and passenger in this text.

Patron Clearance Envelope: The patron reach envelope plus a clearance of three inches.

Pipeline coaster: Riders are positioned between the rails instead of above or below them.

Potential Energy: The energy possessed by an object due to its position. A general equation for potential energy is mass * gravity * height (**PE=mgh**).

Queue: A line you stand in for an attraction, food, or entry/exit.

Racing Coaster: Two separate tracks usually parallel for most of the course. Trains are released simultaneously so they race from start to finish.

Rolling Resistance: A force resisting motion when an object rolls along a surface.

Sit down coaster: Traditional roller coaster with vehicles above the rails.

Spinning coaster: Seats can freely spin on a horizontal axis.

Standup coaster: Riders are restrained in a standing position.

Stacking: When more than one coaster train or vehicle is stopped on consecutive block brakes. Usually means the ride is not being run to maximum efficiency due to slow dispatches or other technical issues.

Statics: The equilibrium mechanics of stationary bodies.

Strain Gauge: A device used to measure the strain of an object by converting force, pressure, tension, or weight into a change in electrical resistance which can then be measured.

Swinging suspended coaster: The vehicle hangs below the rails and can freely swing from side to side but does not invert.

Tachometer: An instrument used to measure the rotation speed of a shaft

Themed/Theming: The central idea or concept for an attraction or area. Conveys the story behind an attraction to the guests.

THRC: Theoretical Hourly Ride Capacity is the number of guests per hour that can experience an attraction under optimal operating conditions. Calculated by: Riders per bench*benches per car*cars per train*(60min/ride time minutes).

Valley: A convex element on a roller coaster's track. Also, the term used when a train or vehicle doesn't have enough momentum to make it over the following element and becomes stuck, or "valleyed." A valleyed vehicle needs to be evacuated and then towed over the next element.

Viscosity: The measure of how resistant to flow a liquid is.

Wing coaster: The seats are fixed on both sides of the vehicle outside of the rails. Intamin calls their lone installation a "wingrider" coaster.

Appendix III: Notable Milestones in Roller Coaster History

1400 - First known roller coaster

1600 - Russian ice slides

1784 - First coaster on wheels built in the Gardens of Oreinabum in St. Petersburg, Russia.

1817 - First coasters with cars that locked to the track. The Aerial Walk became the first full circuit coaster.

1846 – The Centrifugal Railway, Frascati Garden, Paris was the first looping coaster (though it was not a full circuit).

1848 – La Marcus Thompson was born and is commonly referred to as "the Father of the Roller Coaster."

1873 – Mauch Chunk Railway was America's first roller coaster. Built in Pennsylvania, it was the second most visited attraction in America behind Niagara Falls.

1884 – Thompson's Switchback Railway opened at Coney Island in Brooklyn, New York.

1885 - First use of a powered chain lift was on a switchback railway in San Francisco, USA.

1907 - Drop-The-Dips became the first coaster to use lap bar restraints to safely secure passengers.

1920 – The first Golden Age of wooden roller coasters. There were over 2,000 operating coasters during this time.

1925 - Cyclone at Revere Beach, Massachusetts was the first coaster to exceed 100 feet in height.

1959 – Matterhorn opened at Disneyland (Anaheim, California) and was the first tubular steel rail coaster and used a modern control system.

1970 – There were only 172 operating roller coasters in the world at this time, down from over 2000 in the 1920s.

1972 – The Racer at Kings Island sparked the Second Golden Age of Wooden roller coasters.

1975 - Corkscrew at Knott's Berry Farm became the first modern looping roller coaster (relocated to Silverwood in 1990).
1976 - First coaster with a modern (clothioid) loop was Revolution at Six Flags Magic Mountain, Valencia California.
1977 – The first launch coaster by means of a weight drop, King Kobra, opened at Kings Dominion in Doswell, Virginia.
1979 – The Beast opened at Kings Island and to this day is still the world's longest wooden roller coaster (7,361 feet).
1981 – The first swinging suspended roller coaster, The Bat opened at Kings Island in Mason, Ohio.
1982 – The first coaster to use stand-up trains opened in Japan.
1989 - Magnum Xl-200 at Cedar Point in Sandusky, Ohio became the first full-circuit roller coaster to break the 200-foot barrier.
1992 - First inverted coaster was Batman the Ride at Six Flags Great America in Gurnee, Illinois.
1996 – Flight of Fear at Kings Island was the first to use linear induction motors.
1999 – Hop Hari in San Paolo, Brazil opened South America's first wooden roller coaster. Superman: Ride of Steel at Six Flags Darien Lake was the first coaster to use magnetic braking.
2000 – Millennium Force at Cedar Point in Sandusky, Ohio became the first full circuit roller coaster to stand over 300 feet tall. It also was the first modern coaster to utilize a cable lift. Steel Dragon 2000 (Nagashima Spa Land in Nagashim, Japan) opened and is currently the world's longest roller coaster (8,133 feet). Son of Beast at Kings Island is the first modern wooden roller coaster to go upside. It also claimed the record for tallest (218 feet) and fastest (78.4mph) wooden roller coaster ever (the ride was closed in 2009 and demolished in 2012).
2001 – First compressed air launched coaster was Hypersonic XLC at Kings Dominion.
2001 - First 4D coaster, X, opened at Six Flags Magic Mountain.
2002 - Xcelerator at Knotts Berry Farm was the world's first hydraulic launch coaster. Colossus (Thorpe Park in Chertsey,

England) was the first roller coaster to feature ten inversions. The Lost Coaster of Superstition Mountain at Indiana Beach is the first wood coaster to use magnetic brakes.

2003 – Top Thrill Dragster, also found at Cedar Point, was the first full-circuit coaster to stand over 400 feet tall.

2005 - Kingda Ka at Six Flags Great Adventure in Jackson, New Jersey opens as the world's tallest roller coaster at 456 feet.

2010 – Formula Rossa at Ferrari World in Abu Dhabi becomes the world's fastest roller coaster at 149.1 mph.

2013 –The Smiler at Alton Towers in the United Kingdom takes the record for most inversions on a coaster with 14. Outlaw Run at Silver Dollar City takes the record for most inversions on a wood coaster with three. The first modern wooden bobsled coaster, Flying Turns, opens at Knoebels in Elysburg, PA. Last known used of LIMs on a major new coaster.

2014 – Banshee at Kings Island becomes the world's longest inverted coaster. Goliath at Six Flags Great America becomes the tallest and fastest operating wood coaster.

2015 – Thunderbird at Holiday World is America's first launched wing coaster. Cannibal at Lagoon Park uses an elevator lift followed by the world's steepest drop. ZDT's Switchback will be the first modern wooden shuttle coaster.

2017 – Sky Dragster, the first Maurer Rides Spike coaster, opens.

2018 – The first Raptor single rail coasters by Rocky Mountain Construction open.

2019 – Steel Curtain at Kennywood boasts the world's tallest inversion as well as the most inversions on a coaster in North America.

Present – Currently, there are more than 5,000 roller coasters operating worldwide. Eight wooden roller coasters go upside-down (half are outside the US) and seven steel coasters in North America have a drop over three hundred feet.

Appendix IV: Resources

For additional information, please check out the following tools and resources. This book wouldn't have been possible without them.

Books

Coker, Robert. Roller Coasters: A Thrill Seeker's Guide to the Ultimate Scream Machines. Sterling Publishing Co., Inc. 2002

Gregg, Paul. Backyard Roller Coasters: Research and Development.

Munch, Richard. Harry G. Traver: Legends of Terror. Amusement Park Books. 1982.

O'Brien, Tim. Dick Kinzel: Roller Coaster King of Cedar Point Amusement Park. Casa Flamingo Literary Arts. 2015.

Ponstingle, Evan. Kings Island: A Ride Through Time. Rivershore Creative. 2021.

Rutherford, Scott. The American Roller Coaster. MBI. 2000.

Unterman, Nathan. Amusement Park Physics: A Teacher's Guide. J Weston Walch. 2nd Edition. 2001.

Wardley, John. Creating My Own Nemesis: The autobiography of the man who designed Alton Tower's big rides, and brought the Theme Park to Britain. CreateSpace Independent Publishing Platform. 2013.

Wilson, Neil. Absolute Efficiency: A Guide to Operational Efficiency in the Theme Park Industry. Theme Park Press. 2021.

Younger, David. Theme Park Design & The Art of Themed Entertainment. David Younger. 2016.

Professional Organizations

American Coaster Enthusiasts (ACE): http://www.aceonline.org/

Amusement Industry Manufacturers and Suppliers (AIMS) http://www.aimsintl.org/

ASTM International: http://www.ASTM.org

Council for Amusement and Recreational Equipment Safety (CARES): http://caresofficials.org/

International Association of Amusement Parks and Aquariums (IAAPA) http://www.IAAPA.org

Themed Entertainment Association (TEA) http://www.teaconnect.org/

United States Patent and Trademark Office: http://patft.uspto.gov/

Student Organizations and Opportunities

http://cornelltpeg.wix.com/tpegcornell
http://www.psuthemeparkengineering.com/
http://www.themeparkeng.org.ohio-state.edu/
https://disneyimaginations.com/
https://thrilllab.blog.torontomu.ca/
https://www.theparksman.com/college/
https://skylineattractions.com/about/skynext/

Websites

Coaster101: http://ww.Coaster101.com
LinkedIn: http://ww.linkedin.com
Roller Coaster Database: http://ww.RCDB.com
Ride Accidents: http://www.rideaccidents.com
Roller Coaster Physics
http://www.Rollercoasterphysics.wordpress.com/history
Ride Sims: http://www.ridesims.com

Articles

https://www.dailymail.co.uk/health/article-2155276/Student-20-hours-death-developing-huge-brain-tumour-survives-thanks-ROLLERCOASTER-RIDE.html
http://news.bbc.co.uk/2/hi/uk_news/england/gloucestershire/4878396.stm
https://www.thesun.co.uk/news/8901865/womans-life-saved-after-a-rollercoaster-ride/
https://www.flowmotion.nl/UK/applications/vekoma.htm
http://physics.gu.se/LISEBERG/eng/loop_pe.html
https://www.ultimaterollercoaster.com/forums/roller-coasters-theme-parks/32809
https://www.reddit.com/r/rollercoasters/comments/xszbco/hi_were_rmc_ask_us_anything_other/

Roller Coaster Related Companies

      Roller coaster creation is a huge team effort, but one generic person or company often gets the credit. Listed below are some of the well-known and not-so well-known companies involved in the creation of the thrill rides we all know and love.

B&M – http://www.bolliger-mabillard.com
Baynum - https://baynumpainting.com/

Birket Engineering – http://www.birket.com/
Chance Morgan/Chance Rides - http://www.chancerides.com
Consign – http://www.consignllc.com
Emis Electrics - http://www.emis-electrics.de/
Great Coasters International - https://greatcoasters.com/
Intamin - http://www.intaminworldwide.com
Intrasys - https://intrasys-gmbh.com/english/
Irvine Ondrey Engineering - http://irvineondrey.com/
Maclan - https://maclan.com/urethane/roller-coaster-wheels/
Mack Rides – http://www.mack-rides.com
Rocky Mountain Construction – http://www.rockymtnconstruction.com/
Skyline Attractions - https://skylineattractions.com/
The Gravity Group - https://www.thegravitygroup.com
Vekoma - http://www.vekoma.com/
Uremet – http://www.uremet.com

Software
3DS Max – http://www.autodesk.com/products/3ds-max/overview
AutoCAD – http://www.autodesk.com/education/free-software/autocad
AutoDesk: http://www.autodesk.com/
CATIA V5: http://www.3ds.com
DraftSight: http://www.3ds.com/products-services/draftsight/download-draftsight/
EnginSoft: http://www.enginsoft.com/technologies/civil-engineering/structural-roller-coasters.html
SolidWorks: http://www.solidworks.com/
Solid Edge: http://www.plm.automation.siemens.com/en_us/products/velocity/solidedge/
Maple: http://www.maplesoft.com/products/maple/
NoLimits Coaster Simulator: http://www.nolimitscoaster.com/
Rhino3D: http://www.rhino3d.com/
RISA 3D: http://www.risatech.com/p_risa3d.html

Photography Credits

Cover layout and design by John Stevenson
Hades 360 by John Stevenson
Thunderhead by John Stevenson
Raging Bull trim brake by Larry Treece

3D renderings and models by Patrick McGarvey

Photographs by Nick Weisenberger:
Phoenix at Knoebels
Ravine Flyer II at Waldameer
Behemoth vs Minebuster at Canada's Wonderland
Storm Chaser at Kentucky Kingdom
Yukon Striker at Canada's Wonderland
Orion and Diamondback at Kings Island
Time Warp at Canada's Wonderland
Wicked at Lagoon Park
The Racer at Kings Island
Millennium Force at Cedar Point
Tidal Twist at Columbus Zoo and Aquarium
Hollywood Rip Ride Rockit Universal Studios Florida

Diagrams and renderings by Nick Weisenberger:
Track construction techniques
G force directions
PE vs KE vs TE
Speed envelope
Millennium Force free fall example
Vehicle roll rate
Universal's boom coaster

Public domain images:
Promenades Aeriennes, Jardin Baujon (roller coaster at the Folie Beaujon) Paris, c. 1820
Illustration of drag coefficients by Johan Gustafsson
Anton Schwarzkopf's patent for the Shuttle Loop roller coaster

Creative Commons Images Used: Some images were purchased under the license of royalty free stock photography websites. As part of this license, these images cannot be shared, formatted, or modified in any way. Other images are included as part of the Creative Commons License. These sites have been included with full attribution. http://creativecommons.org/licenses/by-sa/3.0

Anton Schwarzkoph [Public domain], via Wikimedia Commons
Sheikra_Layout_Fertig_(en).svg: Fritz Spitzkohl derivative work: Themeparkgc [CC BY-SA 3.0)], via Wikimedia Commons
The original uploader was Themeparkgc at English Wikipedia [Public domain], via Wikimedia Commons
Bolliger; Walter, Mabillard; Claude [Public domain], via Wikimedia Commons
erstellt von Benutzer:Krummbein123 (created by Benutzer:Krummbein123) Freigegeben unter GNU-FDL
G-Force Drayton Manor 094 by Jeremy Thompson
Friction by Keta, Pieter Kuiper, penubag, Wikipedia Commons
X2-firstdrop.jpg by WillMcC, Wikipedia Commons
Mardi Gras ship 22-12-2020 top deck.jpg by kees torn, Wikipedia Commons

Made in the USA
Las Vegas, NV
30 December 2023

83718357R00105